Old Buildings, New Designs

For Courtney

The **Architecture Briefs** series takes on a variety of single topics of interest to architecture students and young professionals. Field-specific and technical information are presented in a user-friendly manner along with basic principles of design and construction. The series familiarizes readers with the concepts and technical terms necessary to successfully translate ideas into built form.

Also in this series:
*Architects Draw*
Sue Ferguson Gussow, 978-1-56898-740-8

*Architectural Lighting: Designing with Light and Space*
Hervé Descottes, Cecilia E. Ramos, 978-1-56898-938-9

*Architectural Photography the Digital Way*
Gerry Kopelow, 978-1-56898-697-5

*Building Envelopes: An Integrated Approach*
Jenny Lovell, 978-1-56898-818-4

*Digital Fabrications: Architectural and Material Techniques*
Lisa Iwamoto, 978-1-56898-790-3

*Ethics for Architects: 50 Dilemmas of Professional Practice*
Thomas Fisher, 978-1-56898-946-4

*Model Making*
Megan Werner, 978-1-56898-870-2

*Philosophy for Architects*
Branko Mitrović, 978-1-56898-994-5

Image Credits
All images courtesy of the author unless otherwise noted on image page or as listed: Cover: Philip Vile; 22: Gilly Walker (Creative Commons licensed on Flickr; http://www.pixellated.typepad.com/); 27 (bottom): Ian A. Holton (Creative Commons licensed on Flickr; http://www.flickr.com/people/poeloq/); 59: Brother Randy Greve (Creative Commons licensed on Flickr; http://www.holycrossmonastery.com/); 100 (right): Flickr user _ppo

Published by
Princeton Architectural Press
37 East Seventh Street
New York, New York 10003

For a free catalog of books, call 1.800.722.6657.
Visit our website at www.papress.com.

Editor: Megan Carey
Designer: Jan Haux

Special thanks to: Bree Anne Apperley, Sara Bader, Nicola Bednarek Brower, Janet Behning, Fannie Bushin, Carina Cha, Tom Cho, Penny (Yuen Pik) Chu, Russell Fernandez, Linda Lee, Jennifer Lippert, John Myers, Katharine Myers, Margaret Rogalski, Dan Simon, Andrew Stepanian, Paul Wagner, Joseph Weston, and Deb Wood of Princeton Architectural Press —Kevin C. Lippert, publisher

Library of Congress Cataloging-in-Publication Data
Bloszies, Charles.
Old buildings, new designs : architectural transformations / Charles Bloszies. — 1st ed.
p.  cm. — (Architecture briefs series)
Includes bibliographical references.
ISBN 978-1-61689-035-3 (alk. paper)
1. Architecture—Aesthetics. 2. Buildings—Repair and reconstruction. 3. Buildings—Additions. I. Title. II. Title: Architectural transformations.
NA2500.B575 2012
720.28′6—dc22
                                                    2011015979

# Old Buildings, New Designs

## Architectural Transformations

**Charles Bloszies**

Foreword by Hugh Hardy

Princeton Architectural Press,
New York

Table of Contents

Chapter 5
# Case Studies

# Foreword

Hugh Hardy, FAIA

This is an important book for anyone concerned about the future of architecture. Our media culture presents buildings from the perspective that their importance comes from being new, with a preference for works that are unlike anything seen before. Instead, Charles Bloszies has put together an incisive and broad investigation of nineteen projects from all over the world that show thoughtful ways in which new buildings draw their importance from their relation to old. His is a generous survey of an extensive and diverse range of design possibilities. More than a handsome publication of designs, however, this book is also a considered exploration of a subject vital to the profession, one that should encourage great discussion.

Bloszies's text investigates why old buildings appeal so strongly to the public and the resulting challenge this represents to contemporary design. Conflicts between public policy concerning landmarks law and sustainable design are also factors he explores with clarity. No professional jargon mars his prose, and his goal "to explore successful design approaches for visible interaction between new and old" is admirably realized.

Architecture only exists once it is built. Future architecture must be spoken of with drawings, images, or digital information. Buildings, however, can occupy the present and also speak to us about the past. Although apparently static, their uses change, and their activities ensure they never remain new. Some even become so redolent with historical content they are made into museums and deliberately kept the same through detailed preservation. As an aesthetic ruse, some are even added onto in imitation of the original. This book assumes a more vigorous premise, using change as the vehicle to introduce new ideas, new ways of building. Furthermore, Bloszies cites how this can be accomplished without bowdlerizing the original, careful to respect the values of those who built it.

By not being doctrinaire, a wide selection of examples instructs how to appreciate each project on its own merits. So much of contemporary architecture is created and judged as a standalone consumer product, but pursuit of these pages proves there is no single way to responsibly shape new in relation to old. Although it becomes obvious that the best old work incites the best new, no two projects offer the same aesthetic proposition, nor should they, claims the author. Instead, he argues for an exploration of ideas that celebrates continuity.

Bloszies provides valuable commentary on why the public's response to historic preservation has been so intense and why it has caused modernism's once tentative embrace of the public imagination to all but disappear. Today that revolutionary aesthetic often lies buried under a tidal wave of moldings, small-paned windows, sloping shingle roofs, and "vintage details" that provide only an ill-proportioned simulation of decorative skill. Instead, this book puts forward a bracing approach to how new can meet old, always assuming a degree of contrast. It cites examples in three categories: *extreme*, *restrained*, and *referential*. Each results from a consistent and carefully realized design premise chosen for its clarity.

It could be argued that the majority of examples shown are small, and it is therefore not readily apparent how these ideas can be applied to large-scale structures. But I suggest this is the virtue of a stimulating investigation. The success of these small-scale efforts can only encourage thinking about how the same ideas could be applied to the increasingly big structures required in contemporary construction. Financing, building codes, development practices, population growth, and greater building density are leading us to a more urban environment. No matter how one approaches the subject of an appropriate response to this phenomenon, the relation of new to old should be a paramount concern for us all.

Bloszies's book presents a sufficiently challenging display of projects to shake up traditionalist thinking and stimulate those who prefer to avoid the problem by using clichés.

This is an important book for the public and professionals alike.

# Preface

This book will explore the union of new and old architecture. Alterations and additions to existing buildings are commonplace, yet very little media attention is devoted to this topic—perhaps because new work is oftentimes designed to blend in with the old in order to avoid controversy. This approach typically leads to an overall banal design, or worse yet, a design lacking in integrity. Could guiding principles be found that would lead to the creation of successful designs in this common architectural genre?

I have asked myself this question many times, especially when clients have sought out our firm to design a major addition to an existing building or to make extensive changes to a significant structure. Where are the precedents, and what can we learn from them? This book started as a quest to find these precedents; what we discovered along the way were numerous examples of exemplary work—both small and large projects by firms of all sizes, with varying degrees of name recognition.

Projects merging new and old are not easy to execute, especially if the existing building is deemed a historic resource. While students of architecture and many practicing architects employ the latest design software to create forms that eschew historic styles, preservationists are digging in their heels to resist change, especially change that might compromise the historic integrity of a traditional building or replace it with an avant-garde structure. Given these widely differing points of departure, is it possible for an architect to create meaningful work when charged with the task of fusing new into (or onto) old?

An incident I witnessed in Italy while on a family vacation gave me insight into this question and ultimately led to the writing of this book. We were taking in the tourist highlights in Florence when a bright yellow Lamborghini zoomed into the Piazza della Repubblica. The car was driven by a fashionable young man, who was accompanied by an equally

A classic style that is replicated in a modern material and punctured by clean, contemporary openings creates a compelling juxtaposition of new and old. Hôtel Fouquet's Barrière, designed by Édouard François, Paris, France, 2006

Courtesy of Édouard François

fashionable young woman. The difference between the shiny, well-engineered machine and the rusticated architecture of the buildings surrounding the piazza could not have been greater. This sharp contrast caused me to marvel over the sophisticated engineering behind the design of the Lamborghini as well as the incredible construction of the old buildings. As I was lost in this enhanced appreciation of both new and old, the young man hopped into the car, revved up the engine to impress his girlfriend, and let out the clutch too fast—stalling the machine! Instantly, the power of the metaphor dissipated. The Lamborghini was no longer a feat of engineering design—it was instead a flashy, overpriced, silly car that soothed the vanity of the obviously wealthy dilettante. It was probably leaking oil on the sacred piazza to boot!

More often than not, architectural interventions are caught between these seemingly oppositional forces. The sleek Lamborghini juxtaposed against the rusticated facade represents a valid design proposition for how new and old can interact. Yet, the fear that a good design will be stalled by critics or approval agencies often overrides the willingness of most owners to take on more risk in order to get a great project completed.

For such a design to be successful, there must be a recognizable degree of contrast between new and old. It need not be extreme—differentiation is the key. Inspired by the encounter in the piazza, I looked for juxtapositions of new and old during the remainder of our trip to Italy: they were everywhere, represented by a Smart car parked in front of a simple old building or a new window cut into the arches of an aging wall.

As the case studies exemplifying this juxtaposition were assembled, a common thread among them emerged. We found that a successful project required not only a well-conceived new design but

The contrast between the Smart car and the old buildings leads to a heightened appreciation for the design qualities of both.

The jarring result of new imprinted onto old calls attention to the building and craft techniques of different times.

also a well-conceived old design. It is much easier to create a counterpoint to an outstanding old building than to a mediocre old building. The general public, however, tends to prefer any existing structure, whether well designed or poorly designed, to new architecture. Why is this? Chapter 1 begins with an exploration of this question. In chapter 2, I posit that cities will experience an increase in development of the existing building stock, in part for economic reasons as the goals of preservationists and sustainability advocates align; some architects will need to adjust their work habits to address this. In chapter 3, I present theoretical arguments for what constitute viable design propositions—based on the prerequisite of contrast between new and old. Projects that include a union of new and old are more difficult to design, gain approvals for, and actually build than entirely new construction. Chapter 4 delves into the execution of these hybrid buildings. The last half of the book is devoted to case studies of exemplary work—a diverse compendium of projects chosen to illustrate that both meaningful and thought-provoking architecture can arise as a blend of new and old.

# Old Buildings

The patina of Notre Dame, acquired over hundreds of years, is part of the building's appeal.

Many old civic structures attract favor because they are familiar. U.S. Capitol Building, Washington, DC

## Appeal of Old Buildings

Old buildings are architecture's comfort food. They conjure up nostalgic feelings and remind us of seemingly simpler times. Their composition and massing are easy to understand, and their familiar ornamentation adds a richness of texture often absent in modernist architecture. For these reasons and more, old buildings are not threatening; people like them.

The very fact that a building is old can contribute to its appeal, as illustrated by Victor Hugo's characterization of Notre Dame Cathedral: "Time added to the cathedral more than it took away. Time spread over her face that dark gray patina which gives to very old monuments their season of beauty."[1] There is no substitute for time. Notre Dame acquired its patina over hundreds of years, eroding away crisp masonry edges and allowing tinges of moss to grow. In many cases, time has been kind to architecture.

Another reason we like old buildings is because they are familiar. Many are well-known civic landmarks, where important events take place daily, or monuments that mark significant historic milestones. They represent social stability and institutions that people can trust. Some old buildings have acquired stature simply due to the fact that they have survived longer than their contemporaries, in part because they are exemplary representations of their frequently recognizable architectural idioms. As a consequence, they have been actively used and well-maintained. Since many of these buildings are public institutions, people have passed through their doors many times, often for important personal milestones. Memories of these events evoke a certain attachment for these venerated monuments.

## Familiar Idioms

Many old buildings still in operation today are examples of architectural styles rooted in classical design principles. In Western cultures, classical architecture derived from Greek and Roman arche-types has endured for centuries, having been revived a number of times—including during the European Renaissance and American Beaux-Arts movements. Every major Western city (as well as many non-Western cities) contains important buildings based on the paradigms of classical design. It is remarkable that the design principles behind so many existing structures can be traced to stylistic ideas that emerged over two thousand years ago and that have changed little since then.

Based on a formal balance, and achieved by adhering to the established rules of composition, the classical idiom is ubiquitous in civic architecture. This style has yielded many familiar, eye-pleasing monuments, and it is not surprising that the general public thinks fondly of buildings designed in this manner. There are architects in practice today (albeit few) who accept classical precepts as a priori truth and who are passionate advocates for classical design theory.[2]

Although architectural movements deviating from classical methodology have gained a foothold at several points throughout history, for the most part, architectural style was defined by formal composition until the emergence of modernism in the early twentieth century. With the advent of modernist thinking, painters, sculptors, and architects began to explore abstract ideas regarding the definition of space. Building forms that emerged from this thinking were so different from the familiar classical imprints that, when they first appeared, only the academic elite seemed to understand and appreciate them.

Modernists had a disdain for ornamentation, as was most vehemently argued by Adolf Loos in his 1908 essay "Ornament and Crime."[3] In this treatise, Loos maintains that the application of ornament is unnecessary and merely embodies fashion that will go quickly out of style. Although many architects today agree with Loos, modernist buildings completely devoid of ornament are rarely embraced by the general populace. Current architectural thinking has transcended modernism by light-years. Loos's famous essay seems easy to follow when compared to the intangible and abstruse precepts that have been added to the architectural lexicon in the past few decades. The emerging avant-garde architects of the twenty-first century have embraced the digital tools that allow the building blocks of sophisticated computer modeling software to be combined in ways that defy categorization into any particular architectural style. The resulting work is compelling and often sculpturally stunning but far removed from the familiar language of classical style both in appearance and in its theoretical starting point.

Computers have given engineers the ability to analyze and model the behavior of almost any form an architect can imagine, and contractors have been able to devise the means and methods to build these forms. Theoretical and technical advancements have contributed to the creation of conceptually complex architecture, some of it purposely fraught with aesthetic contradiction.[4]

It could be argued that a kind of artistic illiteracy prevents the uneducated eye from embracing modern and contemporary forms. Many now-famous, universally adored structures were not initially accepted because their designs were radical departures from the accepted norm of the time. For example, the Eiffel Tower, originally built as a temporary structure, was proclaimed an eyesore by many, but has now become the very symbol of French culture. Nonetheless, it is important to acknowledge the strong emotional attachments many people have to old buildings. The strength of this attachment is in part due to a reaction against forms they do not find appealing and theories they have a difficult time understanding or that are not clearly expressed.

## The Urge to Preserve

By definition, most old buildings truly are irreplaceable; this gives them a special, endangered status in the eyes of many individuals. The desire to save these buildings can be based on rational thinking, emotional dogma, or some combination of the two. The motivation to preserve them can also stem from a personal tie or a fear that a new structure will be inferior to the existing old building.

Still, others are driven to save old buildings in the interest of preserving meaningful architectural qualities. Handcrafted exterior and interior elements, large operable windows, access to natural light, and high ceilings characterize many old buildings and are less often found in new ones. Retention of these details is almost universally desirable.

The preservation urge can be roused, however, by issues unrelated to the architectural qualities of a building. For example, as an emotional reaction against modern architecture, some extreme preservationists have taken the stance that an old building should never be replaced with a new one. This dogmatic approach has caused many local jurisdictions to enact strict antidemolition ordinances that require exhaustive study and review to determine if an existing structure can be razed. A few jurisdictions have gone so far as to declare any building over fifty years old to be de facto historic. Ironically, under these ordinances, buildings based on modernist tenets (a movement that had disdain for historic preservation) become subjects of preservation themselves.

The desire to preserve is sometimes sparked by a distrust of the quality of today's construction. Craft plays a role here. Handwrought features, which modern, unadorned surfaces can lack, invoke

Architectural qualities not typical in new buildings, such as ornamentation, contribute to the desire for preservation.

Under some of the strictest policies, any building over fifty years old (including a modernist structure such as the Lakeshore Drive Apartments designed by Ludwig Mies van der Rohe) is considered historic.

warm, nostalgic feelings. Additionally, some individuals consider work crafted by human hands to be inherently better than the same work fabricated by a machine; this can lead to the false belief that all older buildings are built better than modern ones. While it is likely true that historic structures that are still standing today were built and maintained better than their contemporaries that are no longer standing, advances in construction techniques have, for the most part, led to the creation of more reliable, efficient, and durable buildings.

Since the preservation movement began in the 1960s, architects have advocated the retention of buildings through restoration, rehabilitation, and adaptive reuse. More recently, however, recycling of entire buildings has become recognized as an important cornerstone of a sustainable approach to urban development. Cities worldwide have adopted policies that encourage or mandate reuse of existing building fabric. In chapter 2, this topic will be explored in greater detail.

Unfortunately, preservation concerns have sometimes been used as a means to block new, oftentimes denser, development. In this case, preservation is not linked to the merits of the old structure in place on the site, rather it is employed as a political tool to oppose a project that may have a perceived undesirable consequence, such as increased traffic congestion or blockage of views.

Given the rational, nonrational, and sometimes irrational views on the preservation of old buildings, perhaps it is not surprising that there is rarely consensus about what to preserve and how to do it, especially when the introduction of new architectural elements is necessary. The debate, however, should take into account the myriad technical problems with which old buildings are burdened and should acknowledge that not every building merits preservation.

## Buildings Have Finite Useful Lives

Although the patina of time can indeed improve upon the appearance of a stately edifice, the interior workings of a building are often significantly compromised with the passing of time—and were rarely built to meet the demands of today's world. Buildings are not constructed to last forever and must be regularly maintained to survive their useful life expectancy, usually no more than one hundred years. Building systems (structural, mechanical, electrical, plumbing, fire-protection, security, and communications) eventually become obsolete. Building codes are revised at a pace that typically renders a structure out-of-compliance within a decade. Furthermore, legislated public policy can lead to mandates for difficult and costly physical upgrades to old buildings.

Many buildings over one hundred years old originally had no electricity, no central heating, and no air-conditioning. Structures surviving today that were built without these amenities have had to be retrofitted with new systems to avoid becoming obsolete, unfit for habitation, or dangerous. The mechanical or electrical retrofit of a large building is a significant undertaking: it is often more economical to replace a building entirely than to make such modifications, especially if the updated design respects the original building fabric.

Building fires have brought about significant changes in building codes over the past century. Old buildings typically do not have enough fire exits to comply with current codes and often lack fire sprinklers, one of the most effective flame suppressants. Modern high-rise buildings (by code definition, those taller than about six stories) require enclosed, pressurized fire stairs and vestibules that prevent smoke from filling occupied spaces, a fully automated and electronically monitored fire-sprinkler system, an emergency power generator, a fire pump, and an on-site water tank with sufficient capacity to fight a major fire.[5] Most buildings over fifty years old have none of these features.

Furthermore, one of the most common and charming features of a large, old building is an open, monumental stair, connecting all floors and oftentimes discharging directly into the elevator lobby. Unless mitigated mechanically, this spatial arrangement can lead to a situation where the most recognizable exit is filled with smoke during an emergency.

Many major cities have enacted high-rise life-safety retrofit ordinances that mandate compliance with most of the current fire code requirements. While most states and many countries have a historic building code that will allow certain exceptions to these code mandates—provided the exceptions are deemed not overtly life-threatening—the retrofit of an existing structure to meet fire code standards, especially a code-defined high-rise building, is complex and expensive.

Another significant difficulty presented by old buildings is that they are inefficient consumers of energy. The facades of these structures do include attributes that sustainability advocates would consider good practice, such as operable windows and good solar orientation, which the uniformly articulated and sealed envelopes of many modern buildings do not offer. On balance, however, old buildings do not use energy efficiently. Thermal insulation is usually absent, inefficient single-pane glass is common, and old heating and cooling systems consume much more energy than do modern systems.

If adequately maintained, old structures rarely collapse under normal gravity loads—not surprisingly, since their structural systems are tested as soon as they are put into service. However, in cases of a hurricane or earthquake, older buildings suffer more than newer ones whose designs are based on a more advanced understanding of these phenomena. Active seismic zones are especially troublesome for old structures, which are at risk of sustaining significant damage or experiencing total failure during an earthquake.

Social legislation can also profoundly impact buildings. The most significant example is the initiative for worldwide accessibility for persons with disabilities. The Americans with Disabilities Act (ADA) was enacted in 1990, and the architectural consequences of this legislation have been widespread. Whereas new structures can be designed to accommodate the provisions of the ADA, existing buildings often require significant alterations. Oftentimes, a monumental stair at a building's entrance prohibits access right at the front door. Few architectural designs to correct this particular problem have been successful.

The useful life of an existing building can certainly be extended by addressing the concerns discussed above. After all, a viable design solution to these technical problems can almost always be found. However, the implementation of these solutions can be costly and can easily compromise the architectural integrity of the original design. Since a retrofit generally includes the addition of new visible components, the details of how new meets old are the primary drivers of the success of the overall work.

## Not Everything Old Is Good

Old buildings may be architecture's comfort food, but even those who truly appreciate these structures seek diversity in their architectural diet. Bricks and mortar are static, but styles change as social attitudes evolve. Change for the sake of fashion without regard to history is not desirable, but a proper balance between preservation and change *is* desirable—and difficult to find.

One reason an architectural preservation movement was formed in the United States was

Many who truly appreciate old buildings also seek architectural diversity.

Photographer: Gilly Walker

as a reaction to a mind-set fully formed during the 1960s: anything old was interpreted as old-fashioned and a stumbling block to progress. Who needs old buildings? The backlash that formed to challenge this attitude swung the pendulum in the opposite direction and ushered in a historic preservation movement that has been responsible for saving many irreplaceable architectural treasures. Some architects fear that the pendulum has swung too far in that direction, facilitating the creation of obstructions to worthwhile development. Too often, these architects claim, the requirement for the preservation of an unworthy building results in an unsatisfactory design compromise or, in the worst case, an abandoned project and an abandoned old building.

The call to demolish an old building may be justified, however, if the structure no longer serves its intended purpose. Well before the roof, floors, and walls wear out, demands placed on a building by its occupants change. Tenants rotate in and out of office buildings, successful institutions outgrow their spatial envelopes, and industries vacate fallow space. Once a structure ceases to serve its intended purpose, the owner must weigh a number of complex options. Replace or alter? is the first question, and the architectural merits of the building can frame the argument one way or the other. The structure's design should be honestly evaluated, free of the emotion that can surround these decisions. Stubborn and unreasonable positions taken by strident preservationists have done as much to undermine legitimate historic preservation as have midnight demolitions of architecturally distinctive buildings.

Though it is clear that not all old buildings are worth preserving, finding consensus on whether a particular structure should be demolished is rare. If the building possesses architectural qualities most

As buildings age, they struggle to provide the same level of comfort and enjoyment that well-designed modern buildings can. Not all are worth preserving.
Photographer: Luca Gorlero

stakeholders agree are worthy of preservation, every effort should be made to save it. Whether an old building is restored following the strictest preservation guidelines or retained as an artistic fragment fused to a new form, the energy expended to build it in the first place has not been squandered. Architectural diversity, a crucial component of a livable city, is enriched by the resulting mixture of old and new.

## Interventions

Once a decision to save an old building has been made, the work can take a variety of forms, depending on the number and extent of deficiencies that need to be addressed. The degree of intervention will also depend on the attitude that the designer takes toward preservation of the original architectural fabric.

Most architects would agree that certain, highly distinctive landmark buildings with genuine historic credentials, such as the Parthenon or Independence Hall, should be preserved without modern interventions. Strict interpretation of this attitude leads to freezing the building in time under two possible scenarios: restoring it to its original condition or making no intervention at all, simply arresting the forces of nature that could lead to its ultimate demise. Less significant historic structures are often rehabilitated rather than faithfully restored. Rehabilitation may include a minor change from the original use as well as the introduction of new elements (usually behind the scenes).

Most historic preservation boards in the United States have adopted the "Secretary of the Interior's Standards for Rehabilitation" as a guideline for building restoration and rehabilitation projects.[6] The standards appropriately define guidelines for preservation of significant historic structures. However, many legitimate alterations of old buildings

The Secretary of the Interior's Standards for Rehabilitation

1.
A property shall be used for its historic purpose or be placed in a new use that requires minimal change to the defining characteristics of the building and its site and environment.

2.
The historic character of a property shall be retained and preserved. The removal of historic materials or alteration of features and spaces that characterize a property shall be avoided.

3.
Each property shall be recognized as a physical record of its time, place, and use. Changes that create a false sense of historical development, such as adding conjectural features or architectural elements from other buildings, shall not be undertaken.

4.
Most properties change over time; those changes that have acquired historic significance in their own right shall be retained and preserved.

5.
Distinctive features, finishes, and construction techniques or examples of craftsmanship that characterize a property shall be preserved.

6.
Deteriorated historic features shall be repaired rather than replaced. Where the severity of deterioration requires replacement of a distinctive feature, the new feature shall match the old in design, color, texture, and other visual qualities and, where possible, materials. Replacement of missing features shall be substantiated by documentary, physical, or pictorial evidence.

7.
Chemical or physical treatments, such as sandblasting, that cause damage to historic materials shall not be used. The surface cleaning of structures, if appropriate, shall be undertaken using the gentlest means possible.

8.
Significant archeological resources affected by a project shall be protected and preserved. If such resources must be disturbed, mitigation measures shall be undertaken.

9.
New additions, exterior alterations, or related new construction shall not destroy historic materials that characterize the property. The new work shall be differentiated from the old and shall be compatible with the massing, scale, and architectural features to protect the historic integrity of the property and its environment.

10.
New additions and adjacent or related new construction shall be undertaken in such a manner that if removed in the future, the essential form and integrity of the historic property and its environment would be unimpaired.

The Secretary of the Interior's Standards for Rehabilitation provide guidelines for building restoration and rehabilitation projects.

do not have strict historic preservation as a goal. Changing the use of an old building from commercial to residential is a good example. The success of a residential conversion often relies on the charm of an old building: large windows, high ceilings, and period facade; but radical changes to floor-plan layouts and building systems are also often necessary. Retaining a fragment of a historic structure, or only its facade, may be anathema to preservationists but may certainly be judged as artistically credible by others. Because they are so broadly written, the standards have been used to argue both sides of the coin.

It is not the intent of this book to pass judgment on the historic preservation movement. It is important to note, however, that preservation sentiments vary widely, proponents have strong feelings bolstered by political and social connections, and projects that include new, visible elements are almost always controversial. This tension has led to many poorly designed compromises, as advocates of opposing viewpoints dig in their philosophical heels.

It *is* the intent of this book to explore successful design approaches for visible interaction between new and old architectural styles, and to illustrate these approaches using case studies of exemplary interventions made to old buildings. These interventions include both interior and exterior alterations, as well as additions fused to the original building. The common thread among the designs selected is a thoughtful and clear vision of how new can interact with old.

A successful design approach demonstrates visible interaction between new and old architectural styles. Addition to the Morgan Library, designed by Renzo Piano Building Workshop, New York, NY, 2006
Photographer: Michel Denancé

# Sustainable Urban Environments

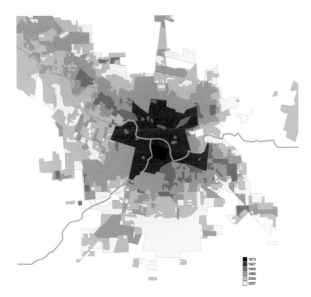

City expansion, resulting from
population increase, often
transforms undeveloped land into
low-density suburbs.
Illustration by Leandro Idecba

Connecting suburban and urban
areas has resulted in all-too-
familiar inefficient transportation
systems.
Photographer: Ian A. Holton

# Critical Components for Urban Sustainability

Cities are nodes of human activity. In developed countries, more than 75 percent of the population lives in urban environments, and these environments are constantly changing in response to social and economic forces. For centuries, change in most cities was driven by an increase in population, resulting in a physical expansion as cities became denser or spilled over, transforming undeveloped land into low-density suburbs. In established urban areas, where inexpensive land on the outskirts was plentiful, growth shifted to the suburbs at the expense of the inner-city core. This trend caused significant loss of vitality within these cities; it reduced natural habitat for wildlife and forced the building of inefficient transportation systems to connect outlying suburban areas to urban centers and to each other. Overall, the nodes of human activity have become less livable as urban growth continues in this manner.

Urban planners began thinking about alternative growth mechanisms in the last few decades of the twentieth century, but it wasn't until the first decade of this century that mainstream planning principles surfaced to address the problems of the prevailing urban growth patterns. Propelled by sustainability advocates, the principles of *smart growth* emerged. The Smart Growth Network defines this movement as primarily influenced by "demographic shifts, a strong environmental ethic, increased fiscal concerns, and more nuanced views on growth."[1]

Demographic shifts can include people moving into cities as immigrants from other countries and suburbanites returning to the urban core in order to experience the cultural benefits cities offer and to avoid long commutes to work. The environmental ethic includes preservation of wilderness and farmlands by discouraging growth on previously undeveloped land and reduction of carbon emissions caused by transportation networks connecting the arms of decentralized sprawl. Smart growth supporters are aware that change in growth patterns is only possible if it takes into account fiscal concerns. For example, the recession of 2008 caused land values to shift; the cost of previously developed sites dropped sufficiently, making redevelopment of inner-city property economically viable. Creative thinking will be required to break the old habits that led to low-density sprawl, like the smart growth movement's nuanced views that encourage high-density, mixed-use developments within existing urban frameworks. Many projects stemming from this thinking will include the retention of existing buildings.

Although not universally embraced, smart growth principles are making their way into public policy. These policies align with historic preservation agendas and generally mandate retention of existing buildings as cultural resources and valuable containers of human activity capable of being recycled for other uses. Smart growth proponents favor increased population density, especially near transportation nodes. These nodes typically coincide with centers of urban commerce and are located in older parts of cities, oftentimes in districts recognized as historic by local or federal agencies. For these areas to become denser, interaction of new designs with old buildings is unavoidable.

Following the precepts of smart growth, old structures, rather than being razed, can be incorporated into future plans for development. Demolition of existing structures in the center of an established city is difficult for a number of reasons: the preservation movement has succeeded in making people aware of architectural heritage, a permit is required in most cities to tear down a building, accompanied by an often cumbersome and expensive review process, and, finally, demolition is noisy, dusty, and typically includes costly abatement of hazardous materials.

As smart growth principles become public policy, and incentives are put in place to encourage an increase in population density, the interaction of new architecture with old will become more commonplace. Linking of architectural styles will also become more prevalent when contemporary additions are made to existing structures. The reuse of old buildings is a critical component of smart growth and will lead to vibrant, diverse, and sustainable urban environments.

## Reuse and Repurposing of Old Buildings

Buildings can become obsolete for a variety of reasons: the original occupants may move to more modern facilities, the structure may be sold to new owners who do not allocate funds necessary for maintenance, or the original utility infrastructure may become too expensive to operate compared to systems typically found in modern construction. Worn-out buildings, however, may not be entirely obsolete, especially those with desirable attributes like high ceilings, operable windows, bountiful natural light, and distinctive facades. Many ordinary old buildings found in urban cores possess these qualities, which are difficult or impossible to replicate. Such structures are ideal candidates for reuse, and many successful small-scale refurbishments have been completed in cities around the world. Other small buildings have been *repurposed*, a term coined by the sustainability movement meaning to find a new use or purpose for an existing entity. In architectural terms, warehouses adapted as art galleries and office buildings converted to residential use are two common examples.

Financial feasibility is usually the driver of reuse and repurposing of old buildings. The 2008

There are many old buildings that have been successfully refurbished and reused in cities around the world.

Photographer: Richard Koshalek

recession affected real estate more than any other sector of the economy. Artificial increases in residential real estate values have been well chronicled, but a similar, less-publicized bubble grew in commercial real estate. Buildings were bought and sold in the decade preceding the recession for record-high prices, leaving the owner little capital available for maintenance and building improvements. As a consequence, the general condition of older structures declined. To make matters worse, as tenants vacated these buildings, the income investors had counted on evaporated, causing commercial foreclosures and lower real estate prices. Investors who were able to hang on to their properties and lenders who acquired buildings after foreclosure sold them well below their most recent purchase amount, at prices similar to those of a decade before.[2]

Real estate values are cyclical, however, and therefore prices should eventually rise. The new, post-bubble owners will have less invested in the purchase price than before, so they will be able to devote capital toward improvements, upgrades, and, perhaps, additions. Tenants interested in smart growth principles and sustainable design in general will be looking for refurbished buildings in the urban core. Since demolition and replacement of old structures is difficult, especially those recognized as historic resources, building owners will likely choose to renovate in order to attract new tenants. Some owners will decide to undertake major reuse and repurposing projects, perhaps stripping an old building to its structural frame and exterior envelope. In cities where unimproved sites are rarely found, owners of underdeveloped land will be tempted to expand structures upward or fill in spaces among existing building segments. A number of the case studies in chapter 5 show how this has been done already.

In a way, reuse and repurposing of old buildings is similar to land reclamation. In the early twentieth century, the U.S. Bureau of Reclamation "reclaimed" overused and eroded land in the western and midwestern regions of the country.[3] The worn-out earth needed to be rejuvenated before it was fertile enough to sustain agricultural activities. Old buildings are kind of like that—they have eroded over time, and the effort required to bring them back to life exceeds that required to build a new structure. Interior finishes, hazardous materials like asbestos and lead paint, and obsolete services must be removed. The exterior envelope must be repaired, and most of this work must be completed before new construction can begin.

The design activities required for a building renovation, especially if significant changes are planned, are also more difficult than those necessary for new construction. The architect must make an assessment of the existing conditions, both technical and aesthetic, before launching into conceptual design work. Unlike the tabula rasa that some architects use as a starting point for a new building design, architects who work on existing buildings must start with a mix of contextual cues, many of which take time and effort to define.[4] Architects typically prepare a feasibility study during a predesign phase so that a potential owner or developer can assess the merits of reclaiming an old building much like civil engineers a century ago assessed the condition of the eroded land and planned for its recovery prior to actual reclamation.

Furthermore, the actual construction of a building that is a union of new and old is more challenging and often more expensive than entirely new construction. Because the condition of the existing fabric that will be retained is difficult to evaluate, construction commences with increased financial risk that unforeseen deterioration and other technical issues might need to be addressed. These issues will be discussed in chapter 4.

Although the economic forces behind smart growth and postrecession real estate values will spur building reclamation opportunities, the additional costs and risks associated with a major reuse project, as compared to a new building project, remain. Public policy, already in place in some cities, will aid the movement to reclaim old, outdated buildings; the following section will outline the complex details urban planners and politicians wrestle with when establishing sound policy.

## Public Policy

Urban land use policy is enacted by city officials based on research by public agencies and private advocacy groups, and sometimes as a reaction to constituent pressure. All major cities have planning departments, responsible for development of broad master plans as well as for approval of individual projects. The master plans lay out policy objectives for the city in general as well as specific strategies for neighborhoods. Once plans have been vetted through a public process, the intentions are written into codes that serve as a framework for regulations to govern land use. Most planning codes are living documents; sections are only overwritten where new policy dictates, and old, sometimes out-of-date sections are left behind. As a consequence, these codes are complex and unwieldy, and the underlying objectives are sometimes difficult to determine.

In some cases, private and public sentiments have changed significantly since the current version of the planning code was ratified, so the laws governing land use do not reflect contemporary community goals. For example, height limits for buildings under the current San Francisco Planning Code reflect a "down-zoning" of the city that occurred in the mid-1980s. At that time, public policy eschewed upward growth, fearing that the character of the city would be erased. As a consequence, in most downtown districts today, the height limit for new buildings is significantly lower than the rooftops of many existing buildings. Today, San Francisco planners and policy makers favor smart growth, but the code presents a stumbling block since the heights of new buildings and vertical additions to existing buildings are paradoxically constrained.

Exceptions in the planning codes add even more complexity to land use policy. In order to break the rules under special circumstances, most codes allow variances from specific requirements and include exceptions that permit conditional uses. The public must agree that variances and exceptions are warranted, and, as a result, projects seeking relief from code constraints must be approved by an appointed body (e.g., a planning commission or a zoning administrator). Once approval of a project enters the public realm, the outcome can be more influenced by political pressure from special interest groups than by the merits of the proposal itself.

Special interest groups, including historic preservationists, have stymied development in urban centers, sometimes for good reasons. All too often, however, arguments for or against a particular project are founded on dogma rather than on a thoughtful examination of the facts. Members of planning commissions are not necessarily design professionals and frequently look for design compromises in order to allow projects with clear public benefits to be approved with minimal political controversy. Projects where new meets old are especially problematic for commissions charged with upholding the public good. One of the aims of this book is to examine case studies of excellent architectural designs that have survived this process.

Quite clearly, cities struggle with the design issues surrounding change. Planning codes in some cities include historic districts where the degree of alteration to an existing building is strictly regulated.

In Venice, alterations of existing buildings are strictly regulated, resulting in historic facades that conceal the modern activities of the inhabitants.

In Times Square, alterations of buildings are much less restricted, resulting in an overlay of modern technological interventions on a mix of new and old structures.

Photographer: Ken Thomas

Other cities designate specific areas where transformation is not so restricted, if at all. Extreme examples are old European cities like Venice, Italy—facades frozen in time, concealing modern activities within—and, on the other hand, Times Square in New York City, which oozes modern technological excitement with an overlay of digital artistry on a mix of old and new buildings. Both examples are important urban environments with clear histories that caused them to develop or not develop in the manner they did. Yet these two extremes are frequently cited to foreshadow the consequences of not allowing change (Venice) or allowing too much of it (Times Square).

The goal of land use public policy is to find the correct balance. Without change, cities will stagnate, but with too much change, urban character will be lost. The best balance leads to cultural diversity—one major reason people live in cities. This concept can be extended to architecture: a mix of new and old—sometimes delicately blended, at other times overtly contrasting—leads to an architecturally diverse urban environment.

## Will Smart Growth Take Place?

Smart growth has already become a policy cornerstone in some cities.[5] Its effects will be seen sooner if policy makers put in place incentives that will cause urban centers to grow denser. Policies will be implemented one building at a time, not altogether, and not overnight. The retention of old buildings, not only historic ones, will be an important element of smart growth; more old structures will be refurbished and, in many cases, expanded—horizontally, if land is available, or vertically, if not.

Smart growth has already taken place in some cities where urban infrastructure improvements, such as efficient transit systems and allocation of open space, have kept pace with private

Dense cities are often the most livable cities.

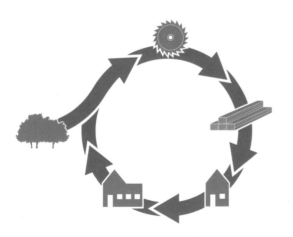

In order for a system to be sustainable, it must replenish the resources it uses.

development, resulting in increased density. Many livable cities today are also dense cities.[6] For smart growth to emerge in other cities, land use regulations will need to change, and incentives will need to be put in place to encourage sustainable development. City officials will need to make developers aware of existing incentives, such as tax credits for preservation of historic buildings, and facilitate their implementation.[7] In some cities, zoning changes will be necessary to allow taller buildings or larger developments with more floor area. Arcane regulations governing the transfer of development rights ("air rights") will need to be adjusted so that vertical additions to existing buildings are allowed.[8]

If properly planned, cities that grow based on smart growth principles will not only be more sustainable but will also be more visually diverse. Smart growth policies will encourage developers to build structures that meld new and old construction; the sustainable design and aesthetic implications will need to be considered by the architects of these projects as outlined below.

## Sustainable Design Implications

A sustainable system endures because it does not deplete the nourishment on which it thrives, in part due to efficient consumption. Similarly, a building is sustainable if it fulfills a lasting need. For example, the Duomo at Siracusa (which will be discussed further in chapter 3) has survived for over 2,500 years and remains viable today. It fulfills a cultural need as the religious seat of the city of Siracusa and has been modified a number of times in the past to maintain this status. In a way, the story of the duomo is an ancient illustration of smart growth principles.

Sustainable design often starts with developing a strategy to minimize the energy required to operate a building, but a broader definition of sustainability includes taking into account the energy required to build it in the first place. Retaining an existing building is far more efficient than demolishing and recycling its components because the energy already expended to build it will not be squandered.

A revision to the LEED energy performance standards illustrates a growing understanding of this point.[9] Under LEED 3.0, the number of credits available for development of an existing site and building reuse has increased from eight credits to nineteen credits, a recognition of the value of the energy already spent to construct the original building.

The LEED point system is only one yardstick used to measure building energy performance—not all sustainability proponents view energy use in buildings the same way. For some, the chief goal is zero energy use, while for others it is obtaining a zero carbon footprint. Some standards include the energy required to construct the building while others do not, since this energy can be amortized over the useful life of the structure. While all of these viewpoints have merit, the most desirable is the quest for carbon neutrality, including the energy required for extracting, fabricating, and assembling the building's components. Using this metric, it is clear that the reuse of as much of an existing structure as possible is an intelligent approach.

However, existing construction typically must be modified to meet current energy-performance standards. For example, the exterior envelope of an existing building may or may not be an efficient thermal insulator. Exterior walls were oftentimes constructed of thick masonry, suitable for heat energy storage, and passive solar control features, such as overhangs and deep-set windows, were commonly employed. But the same structures frequently contain windows glazed with single-pane glass, a poor insulator. Reglazing windows with high-efficiency insulated glass is a common energy

There are many yardsticks to measure sustainability.

upgrade. Overall, although not as energy-efficient as today's highly engineered curtain wall systems, the facades of old structures, when upgraded, do perform well. It is important to weigh all related factors when making a claim that refurbished building fabric requires less energy to put into service than does new construction.

## Aesthetic Implications

Individual buildings have been characterized as stitches in the urban fabric. What will this fabric look like as more stitches are added—a patchwork quilt or a rich tapestry? Clearly, the addition of solar panels or wind turbines on the roof of Notre Dame would cancel out the patina it has acquired over the centuries. In fact, there are few aesthetically successful examples of so-called integrated design, in which energy-generating components are incorporated into the architectural fabric, of new buildings, let alone old structures; the results of such integration often look as though some sort of science experiment is being conducted on the roof.

Perhaps one approach that can lead to a kind of aesthetic sustainability, in which a design can endure well beyond the moment it was created, is for architects to take cues from the time-tested passive features of old buildings that still meet rigorous performance criteria. Some of these features, such as overhangs and sunscreens, may be rendered in a modern architectural vocabulary and applied to old facades or integrated with new fabric as a kind of functional ornament.

Architects of the future will undoubtedly take different paths to achieve aesthetic sustainability, and buildings that are a union of new and old will add complexity to the task. Adjustments to their philosophical positions concerning the influence of context may be necessary, a topic that will be examined in the next chapter.

Integration of energy-generating devices into the visual fabric of new buildings has yet to be achieved. It will be even more difficult in old buildings.
Photographer: Terry Whalebone

Chapter 3

# Design Propositions

## The Question of Context

For centuries, architects have grappled with the question of how a building should relate to its surrounding context. Should it be a singular object, addressing its own set of needs unrelated to its site, or should it blend into its surroundings, as an element of a larger composition?

Throughout the history of architecture, these distinct design approaches have led to an astounding diversity of built work. Architects have passionately argued their viewpoints through their designs and in written manifestos; examples of good design as well as poor design can be found in both categories. When new forms are physically joined to old forms, however, the question of context is more immediate. This architectural fusion overtly exposes differing philosophical perspectives as architects propose individualistic interventions or designs that are seamlessly integrated into the existing urban fabric.

At the turn of the twenty-first century, architectural critics seemed attracted to movements at both ends of the philosophical continuum. Complex, computer-generated, object-oriented designs could be found at one end and high-performance designs striving for a carbon-neutral footprint at the other. Although these positions are not necessarily mutually exclusive, the projects that garnered the most attention made it seem so. The dramatic, nonrectilinear forms designed for the most conspicuous consumers were inefficient by many yardsticks of sustainability, while the most efficient buildings tended to be architecturally timid.

As argued in chapter 2, smart growth principles, established historic preservation precepts, and interest in sustainable design will advance the need for alteration of old buildings. For hybrid buildings of both new and old architecture, is there a particular design philosophy best suited to achieving a sustainable result aesthetically that will withstand the test of time?

This question will be examined below, drawing some general conclusions from the two opposing design approaches posited above. The key aesthetic parameter is how new meets old, and it will be argued that a number of different points of departure, all stemming from a common root, are valid.

## Lessons from History

Only a few notable examples of architectural styles combined in a single building exist historically. New buildings tended to rub out old ones, and those structures that survive are the fittest examples of architectural styles that prevailed when they were built. Styles changed slowly over the course of architectural history, and unlike today, buildings constructed within any particular decade—and in some cases within a given century—are similar in appearance.

Assuming that architectural history commenced around 3,000 BCE, the most enduring style has been Egyptian. As an analogy, if the past five thousand years of architectural history were compressed into a twenty-four-hour day, the Egyptian period would last about ten hours. Classical architecture would consume about seven hours, the Renaissance around an hour, and recent movements like postmodernism would be over in about three minutes.

The pace at which these architectural styles changed historically parallels the rate of civilization's evolution in general. Recent acceleration is mostly due to technical innovation. For example, structural steel and modern reinforced concrete allow today's architects to design almost any shape at almost any scale. On the twenty-four-hour architectural timeline, innovations that completely unbridled architectural creativity have come about in the last few seconds. Today, style designations have little meaning, current works do not fit into an easily recognizable stylistic pattern, and contemporary architectural design attitudes are anything but static.

Nonetheless, the question of context remains, especially when new meets old. Before suggesting a set of design propositions to address this question, it is instructive to look at very old examples of architectural fusion through modern eyes.

The Duomo at Siracusa in Sicily consists of a blend of three architectural elements, spanning more than two thousand years. The oldest part is Greek Doric, a style dating from around 500 BCE. A Romanesque nave and second story most likely replaced the roof of the Doric temple during the Norman period (eleventh century), and a baroque facade and apse were added after an earthquake in 1693. Juxtapositions of these three distinct styles are visible from various vantage points both inside and outside of the cathedral.

Despite the vast time spans between the additions of successive elements, the architectural composition is not fragmented. Perhaps because all three elements are of similar color and were constructed using similar materials, the different eras are not immediately perceived. Furthermore, each of the three portions was appended as a means to the same end—creating the most important religious building in the city. Since the baroque facade is the dominant element, the cathedral looks, at first glance, like a typical Sicilian church constructed during the late Renaissance.

The Mosque at Córdoba, Spain, elicits a different visual response. The site was originally a Christian church, built by the Visigoths around 500 CE, but was replaced by a Muslim mosque around 800 CE. The mosque is rectangular in plan and consists of arcades comprised of brightly colored, double-horseshoe-shaped arches, supported by slender

At the Duomo at Siracusa, three distinct styles—Greek Doric, Romanesque, and baroque—are visible from various vantages of the cathedral.

Photographer: Giovanni Dall'Orto

The architectural makeup of the cathedral feels cohesive despite the disparate styles.

Photographer: Giovanni Dall'Orto

At the Mosque at Córdoba, the two architectural styles are clearly differentiated and pronounced.

The insertion of new into old is blunt—the Gothic fan vaulting collides with the horseshoe arches.

Photographer: Luca Volpi

columns. Córdoba returned to Christian domination in the thirteenth century, and around 1525, a Gothic nave was inserted into the center of the mosque. Architecturally, the insertion is blunt—the horseshoe arches visibly collide with the Gothic fan vaulting. Shafts of daylight pierce the dark composition of the mosque in a dramatic manner, but in the church, daylight illumination is bright and uniform. The contrast between the two architectural styles could not be more pronounced, echoing perhaps the differences between the religious beliefs that prevailed in the sixteenth century.

In both the Duomo at Siracusa and the Mosque at Córdoba, the architects of the newer portions of the structures did not choose from an array of possible styles; they simply used the style that prevailed at the time. The differences between new and old are revealed through honestly expressed contrast, subtle in Siracusa and harsh and overt in Córdoba.

## Design Integrity

Present-day musicians appreciate and perform classical music, but very few modern composers write classical pieces following the exact principles of the Renaissance movements. It is a genre from the past, worthy of preservation but not of emulation. Music is a product of a culture and of a time; therefore, duplication of a past style lacks artistic integrity. This is true of architecture as well, although the desire to see new buildings as clones of old buildings persists (as discussed in chapter 1), at least among a segment of the general population. To some, an addition that mimics the original building is appropriate, creating a seamless amalgamation of new and old.

During the nineteenth century, the issue of what it meant to restore a building was a point of debate between two influential architects, John Ruskin and Eugène-Emmanuel Viollet-le-Duc. Ruskin argued that honest restoration was impossible since the original work and the new construction were separated in time, while Viollet-le-Duc was a proponent of mimicking the original work with new construction. According to Ruskin:

> Neither by the public, nor by those who have the care of public monuments, is the true meaning of the word *restoration* understood. It means the most total destruction which a building can suffer: a destruction out of which no remnants

can be gathered: a destruction accompanied with false description of the thing destroyed. Do not let us deceive ourselves in this important matter; it is *impossible*, as impossible as to raise the dead, to restore anything that has ever been great or beautiful in architecture.[1]

Viollet-le-Duc's view was that a restoration is "[a] means to reestablish [a building] to a finished state, which may in fact never have actually existed at any given time."[2]

Ruskin would not have supported an addition to a building that mimicked its original form or, for that matter, the many restoration techniques that are today endorsed by the most conservative preservationists.[3] Viollet-le-Duc, however, put his philosophy into practice with numerous additions and alterations to existing structures throughout nineteenth-century France, frequently adding building features for which there were no historic precedents.

The most famous (and clear) representation of Viollet-le-Duc's design approach is in Carcassonne, France, where he embellished a crumbling medieval city with new gate towers that had never before existed, but which imitate the style of the surrounding architecture. The resulting composition is an inaccurate, romantic expression of the past. Without special knowledge of medieval history, it is impossible to perceive this architectural anachronism.

Whereas Viollet-le-Duc maintained that new indistinguishable from old is a valid design proposition, Ruskin argued that it is impossible to restore old work faithfully. As a matter of principle, Ruskin's position is undeniably correct. For a design to have integrity, it must be a product of its own time—an honest expression of the cultural forces active when the design was executed. In Ruskin's classical tradition, truth equals beauty.

Mimicking the style of the surrounding architecture, Viollet-le-Duc added gate towers to the crumbling medieval city of Carcassonne, France, creating an inaccurate expression of the past.

Photographer: Jean-Pol Grandmont

With architecture as a notable exception, works of art are rarely altered or expanded purposefully, unless the artist had not completed the original. Buildings, however, are frequently altered and occasionally expanded. It would be tempting to adopt Viollet-le-Duc's approach, imitating the original work, but the result would lack integrity because differentiating between the new and old components would be impossible.[4] A time stamp must be unmistakably identifiable on both parts.

## Contrast

Some degree of contrast is an a priori condition of distinguishing between works executed at different times. As illustrated by the Duomo at Siracusa and the Mosque at Córdoba, contrast may be subtle or overt. Like the Lamborghini in the Piazza della Repubblica described in the book's preface, extreme contrast can cause enhanced appreciation of both new and old.[5] On the other hand, restrained contrast, usually employed to differentiate similar architectural elements, can be equally clear. Contrast in which new is referential to compositional rhythms of the old, although more subtle than extreme or restrained contrast, is yet another effective design approach. Architectural unions that aim to achieve contrast via any one of these strategies (extreme, restrained, or referential) can lead to a valid design proposition.

The designs chosen for the case studies in chapter 5 will illustrate these three approaches. The difference between extreme and restrained contrast is a matter of degree; both tactics will result in a clear distinction between new and old forms. Referential contrast is more subtle because the architectural references between new and old components may not be obvious at first glance.

## Critical Viewpoints

New buildings receive the majority of architectural criticism in the media today.[6] The issue of context is mentioned occasionally; in rare cases, it is used as a yardstick to measure a project's success. Critics generally praise avant-garde, object-oriented work, which by its very definition contrasts with its surroundings, but seem reluctant to comment on context. Yet, context is the single most important issue to those who oppose new, innovative work.

Very few examples of fusions of new architecture to old can be found in the architectural media, and those that have been published are mostly additions to venerable institutions, typically expansions of famous museums. The purpose of this book is not only to examine this architectural convergence of new and old from a theoretical viewpoint but also, more importantly, to show examples of successful designs at various scales where the old component retains its design integrity.

Since many people like old buildings and tend to fear new designs, critics will be challenged to explain why cloning past architecture is not an honest design proposition. There will be many voices of dissent: some practicing architects advocate traditional styles, and preservationists may not favor making any changes to certain buildings. Some individuals may not agree that contrast can lead to an enhanced appreciation of both new and old. Many of the case studies cited in this book were highly controversial projects; in general, the more extreme the contrast between new and old, the greater the controversy.

## Exemplary Work

If architects, academics, and critics were asked to list the attributes of good design, their answers would vary widely. However, most would agree that thoughtfulness and clarity of purpose would rank high. All serious architecture is based on an underlying thesis—the idea behind the work. The resulting design is judged by how successfully the architect executed the design intent.

The case studies in chapter 5 have been chosen to illustrate the theoretical points outlined above. The projects all employ contrast as bases for their designs, and most reveal the architect's deep understanding of the immediate context represented by the existing construction. The best work results when the architect has combined respect for the old with a skilled command of the new.

# Project Execution

Execution of designs for new architecture linked to old differs significantly from the execution of designs for new buildings, starting with agency approvals and culminating in a completed structure. As previously discussed, alterations of historic buildings can be controversial, and construction on dense urban sites is almost always difficult. Construction costs can be hard to control because of unforeseen conditions within the existing building envelope. Furthermore, the number of interested parties who may influence the design is often greater for a major addition to an existing building than for an entirely new structure.

## Stakeholders

The first step in the execution of a building design is to gain approval from the various stakeholders with an interest in the project. Approvals are necessary for new buildings, of course, but approvals for alterations to existing buildings can be more cumbersome. Typically, the stakeholders are:

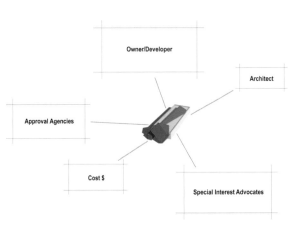

Owner/Developer

Architect

Approval Agencies

Cost $

Special Interest Advocates

Project stakeholders and influences

> Owner/Developer
> Architect (including the entire design team)
> Local Planning Department
> Special Regional Commissions
> Local Building Department
> Special Interest Advocates

The desires of these various stakeholders are oftentimes in conflict. The owner/developer generally wants the best value for the lowest cost.[1] The architect wants to preserve the integrity of the design as it takes real form during construction. Planners want projects to serve the common good, and building departments want buildings to function safely. Special interest advocates, almost by definition, support their own often narrowly focused agendas.

Planners may worry that after a preliminary design has gained the support of various commissions and governmental authorities, it will be modified to reduce costs through a value engineering process, often resulting in an inferior design.[2] In an attempt to ensure that the work will be executed as anticipated, approval agencies will include specific conditions that must be met prior to granting permits to occupy the completed project. For example, if an owner has taken advantage of an incentive for historic preservation (such as a tax credit), precise conditions are recorded, including the design team's detailed specification for the work. Agency inspections take place after completion of construction to determine if the work has been executed according to the specifications. If not, the benefit of the incentive is not given to the owner.

Special interest groups can be both helpful and problematic. Historic preservation advocates are common stakeholders when projects include old buildings, whether or not the existing structures are actually historic. The focus of preservation groups can be beneficial when a significant building is threatened with demolition. In such cases, a successful solution could be a melding of new and old rather than an entirely new building. On the other hand, some preservationists will not be satisfied with anything short of a faithful restoration of an existing structure, which can make approval of an addition cumbersome or unattainable, notwithstanding the merits of a good design proposal.

Successful projects are based on designs that balance the desires of the stakeholders. In general, the public approval process weeds out unreasonable demands by tangential interests, although this can be a time-consuming endeavor. Recommendations made by credible special interest advocates, approval agency staff, and commissioners can improve a design. It is very important, therefore, that the architect understands the public process and is willing to hand off control to other project consultants when issues develop that are unrelated to the merits of the design.

## Expectations

Most owners realize that agency permits are required for construction, but very few accurately plan for how many overlapping jurisdictions will dictate the number and type of approvals required for a specific project and how long the approval process will take. For rehabilitation projects, the final scope of work may be contingent on what additional work is "triggered" for upgrades to existing construction by local codes and ordinances based on the size of the project.

Typically, approvals fall into one of three categories: (1) environmental impact, (2) planning and zoning compliance, and (3) conformance with building codes. The first two categories are often combined and commonly known as entitlements. The entitlements process is followed by government agencies that determine if an owner is allowed to use the land in the manner proposed. Most codes and ordinances are written for new construction; work on an existing structure can cause planners and building officials to impose an additional set of conditions on a project in each of the above categories, but also may allow an owner to take advantage of special incentives or code interpretations that apply only to historic buildings.

All construction will have some degree of environmental impact. For example, in California, under the California Environmental Quality Act (CEQA), all projects must be assessed for environmental impact prior to approval as part of the entitlements process. A site with an existing structure must be evaluated to determine the degree to which the proposal impacts the existing conditions when viewed as a historic cultural resource. Previously developed sites and the buildings on them usually contain toxic or hazardous materials left behind from former uses and out-of-date building practices. These must be abated or mitigated as a condition of approval. Additionally, a change in use or expansion of a building can alter the transportation patterns near the site, and these also must be assessed.

Under CEQA, a project will fall into one of several general categories, depending on the extent of environmental impact. Projects may be approved in any of the categories, but the amount of time it takes to make a final determination varies widely. In San Francisco, for example, it will take three to six months for an environmental planner to determine if a small project is categorically exempt (i.e., CEQA does not even apply). A negative declaration, meaning a project that does not have a significant environmental impact, can take over a year. The assessment to determine that the project does in fact make a significant impact will require drafting of an environmental impact report (EIR), which, along with approvals, may take a number of years to complete. Furthermore, legal challenges to CEQA findings are commonly used by opponents to delay controversial projects, especially ones that include a historic building. Developers must anticipate the oftentimes drawn-out review and approval process as a prerequisite for urban development.

Most cities and counties have planning departments that establish land use policies, resulting in zoning regulations that control the use, height, and bulk of proposed construction. Land use regulations change over time and are subject to political forces. It is not unusual for an existing building to be out of compliance with some aspect of the zoning limits currently enforced for its site. Rehabilitating or adding to an existing noncomplying building may trigger the requirement to correct the existing code

Local politics often influence a project once it has entered the public realm. Website screen capture, *San Francisco Chronicle*, 2001

deficiencies. Major jurisdictions realize that not all land use issues are black and white, so a number of legal vehicles typically exist to address gray areas, subject to a public approval process. Conditional use permits and variances are two of these vehicles, and approval of a project with a conditional use permit and/or variance usually occurs at a public hearing in front of a planning commission or a zoning administrator. In very large cities or jurisdictions with an active preservation community, historic building issues are addressed by a separate historic preservation commission. Furthermore, once a project has entered the public realm, it is subject to the forces of local politics.

Existing buildings are commonly out of compliance with current building codes as well, since structural, fire and life-safety, and disabled-access requirements change as research and experience work their way into building regulations. The new construction must comply with current building codes, but in some cases, noncompliant conditions in the old structure may be left as is if they were legally constructed under the code in place at the time of the original construction. In general, the goal of most building departments is to bring the building stock closer to compliance with current codes when alteration work is proposed. In most jurisdictions, the degree of noncompliance allowed is a matter of project economics. Small projects will not cause building officials to request complete reworking of a building to bring it into absolute compliance. However, building officials will require a greater degree of compliance for large projects since an owner planning to make substantial improvements to a structure should devote resources toward building code related improvements as well. Most codes include language that defines when full compliance is required for the entire project. One or more of the following will usually trigger the requirement for building upgrades throughout to meet the current code:

1. Change in occupancy that results in an increased number of occupants or increased usage (e.g., a conversion from commercial to residential occupancy, where the building is occupied at all times).
2. Substantial scope of work—usually work on more than two-thirds of the floors, not including the basement.
3. A substantial vertical or horizontal addition.

The cost of full compliance can dictate whether a project is financially feasible or not. For example, in active seismic zones, strengthening of an old structure to resist earthquake forces will consume a significant portion of a project's budget.

Egress is the most difficult fire and life-safety issue to address in existing buildings, especially high-rises. As discussed in chapter 1, modern high-rise buildings must have fire suppression and smoke control systems, which typically include fire sprinklers, stair vestibules, an emergency power generator, fire pumps, standpipes, and an on-site water tank. These components rarely exist in historic structures. An effective design strategy to mitigate these conditions is to provide the above components in the new portion of the project rather than attempt to conceal them within the original envelope. Sometimes a sophisticated smoke control model can be created to demonstrate that the original building may not pose a fire and life-safety hazard even though its exact configuration would not be allowed under current codes. Building upgrades to meet disabled-access codes when adding to existing buildings can also be an extensive and costly task. Prior to passage of the Americans with Disabilities Act (ADA) in 1990, buildings did not accommodate persons with disabilities very well. Entrances, elevators, and restrooms are typically out of compliance in structures that were constructed prior to enactment of building code requirements based on the prerequisites of the ADA. For example, to make an elevator compliant, the cab typically needs to be larger; as a result, structural alterations would need to be made to increase the dimensions of the elevator shaft for the full height of the building.

Some states and countries have adopted special building codes for historic buildings. To be eligible for the application of these codes, a building must have historic credentials, which could include individual designation (usually in the jurisdiction's planning code) or inclusion in an officially recognized historic district. Fire and life-safety, structural, and disabled-access concerns usually trump preservation interests, but the intent of the code is to allow historic conditions that do not pose a life-safety hazard to persist unaltered.

As can be imagined, the criteria under which full or partial compliance is required are subject to widely varying interpretations by building officials. Owners often assume that their particular building is exempt from new building code requirements because its condition has been "grandfathered" in, which is not always the case.[3]

Financial incentives are available to owners who rehabilitate historic buildings, applicable to preservation construction activities and sustainability initiatives. In active seismic zones, strengthening an existing structure is viewed as historic preservation because, without the retrofitting, the building could be razed by an earthquake. A core tenet of sustainability is reuse of materials, which can incorporate recycling of an entire structure. Incentives include federal tax credits for historic preservation, reduction in property taxes, and utility rebates. The certification process for capturing any one of these incentives can take time and should be dovetailed into the project design schedule.

Whether the building owner is an institution, a developer, or an individual, the cost of construction is the most difficult aspect of a project to predict and control, especially when the work includes a combination of new and old architecture. Estimating construction costs for a rehabilitation project is much more difficult than for a new building, due to potentially unforeseen conditions within the existing building envelope or below grade. An experienced design team will undertake exploratory investigations early in the design process to ferret out the most likely unknowns, and experienced owners will include a cost contingency in the project budget to account for these variables. For work that includes an existing building, the contingency can be as much as 15 percent of the estimated construction cost. A successful project will exhaust this contingency, since it represents work that the owner, design team, and builder think will actually be needed to execute the design.

## Design Difficulties

Projects that combine new and old building fabric pose technical design and detailing challenges not encountered in entirely new construction. Since part of the final assembly already exists, some owners believe that the design and documentation efforts required to complete the hybrid of new and old should be easier (and consequently design fees should be lower) than for a new building. Actually, the opposite is true.

The existing conditions must be assessed, field-measured, and documented in a preliminary manner before design can commence. Thorough research to obtain original construction documents must be made since many aspects of a building that are hidden from view are revealed in drawings and specifications. If drawings cannot be found, exploratory investigation and material testing should be performed. Building and planning code constraints need to be identified early in the design process; timely mitigation of existing deficiencies can lead to a well-integrated design strategy for the new intervention. Once the preliminary design concepts have been established, additional field investigations are necessary to verify initial assumptions and to prepare a set of existing-conditions documents that will be used to figure out how new construction will meet old.

In addition to the architectural studies outlined above, the design team should carry out a number of engineering studies of the existing construction, much like the analyses that are typically done for a historic building restoration. The condition of the original elements that will remain should be assessed, including the energy efficiency of any portions of the existing envelope that will be retained.

Although it might appear to be a simple task, cleaning the exterior of an existing building is not always straightforward. Harsh chemicals can remove historic fabric in ways that are contrary to accepted historic preservation guidelines; additionally, the patina of time can be unintentionally removed by overcleaning flat surfaces. Mocking up small areas using different methods during the design phase of the project will permit the final cleaning method to be specified prior to construction and will also allow more accurate cost estimates prior to negotiating the construction contract.

Structural concerns unique to construction projects that include existing buildings must also be addressed; shoring, underpinning, and strengthening of existing structural elements are examples of such concerns. Most major renovation projects include removal and replacement of structural components, which generally results in the need to

Cleaning the exterior of an existing building can be an involved and expensive process.

Removal and replacement of structural components generally requires shoring the remaining elements.

shore the remaining elements temporarily. Vertical additions will require new supporting structure threaded through the existing structural system or bearing on existing structural elements. In conversions of industrial buildings to lighter uses, the columns, walls, and foundation system may have enough reserve strength to permit adding a few stories without strengthening the elements below. If strengthening of existing structural components is necessary, a number of time-tested methods can be employed. For example, steel columns can be strengthened by adding a concrete jacket, and brick masonry walls can be reinforced by application of a layer of shotcrete.[4]

Most projects that involve combining new and old construction will require foundation work, which is complex and risky and should be approached conservatively. A number of techniques are available, but all include shoring columns one by one to free the foundation of existing loads, strengthening or replacing the columns, and finally transferring the loads back to the new or enhanced elements. Construction underneath an existing building is sometimes proposed to add below-grade parking or other activities that do not need natural light. When underpinning of an existing foundation is required, small segments of the new foundation system are installed in phases to ensure that the structure is never fully supported on temporary shoring.

An especially difficult task when linking new construction to old can be the development of the joints between new and old finishes, particularly exterior joints that need to resist environmental forces. For example, old masonry walls can be severely weathered, uneven, and out of plumb, making alignment with new walls nearly impossible. The new wall may be attached to a flexible frame, but this makes it necessary for the joint to accommodate structural displacements as well. One effective

Detailing the intersection of new and old structures can be challenging, especially exterior joints that need to resist environmental forces. Detail of the Royal Ontario Museum, designed by Studio Daniel Libeskind, Ontario, 2007

Photographer: Brother Randy Greve

solution is to include a transition element as a sort of connective tissue between old and new. These elements are visible and become part of the architectural vocabulary as illustrated in a number of the case studies.[5]

Only a few of the technical challenges related to designing projects that merge new and old have been mentioned above, but clearly they differ from the issues faced by designers of new buildings. The general contractors who build these types of projects also face unique conditions not presented by new buildings, and projects of this ilk constructed in dense urban settings provide additional challenges still.

## Building in the Already-Built Environment

Construction cranes on the urban skyline have become as ubiquitous as the skyscrapers they have helped erect, but most people, including architects, are unaware of how the cranes themselves are assembled on a construction site. If an existing building is already on the site, especially one that takes up the entire footprint, setting up the crane is much more than a trivial problem.[6] On a constricted urban site, the foundation for the crane can sometimes be larger than the foundation required for the project itself; in these cases, a special footing for the crane must be installed first and then tied to existing foundation elements. These elements are later incorporated into the foundation system for the entire project. A common dilemma for both new construction and large renovation projects is finding a place to lay out the crane prior to setting it in place. It is not unusual to do this work on weekends or at night since it may be necessary to block major streets and dismantle transit lines temporarily.

Most construction projects that blend new and old include selective demolition of components

Deconstructing a building to save some components is much more complex than demolishing an entire structure.

that will later be replaced. Additionally, construction that will remain must be shored and protected from the normal demolition problems of noise, dust, and hazardous materials abatement. Taking apart some pieces of a building while saving others is much more difficult than razing an entire structure.

Growing cities will always be in transition, with layers of construction regularly being subtracted and added. Erecting a crane and preparing the site are necessary for building any large structure, but a close examination of the processes involved reveals that building on an already-built site presents additional demands as described above. Nonetheless, the ingenuity and creativity of builders will continue to permit these obstacles to be overcome.

## Successful Execution

Executing a design without compromising its integrity would appear to be fundamental, yet many built works fall short of the original vision for the project. Financial pressure to reduce construction costs, heightened during times when material prices and labor rates are escalating, is often the culprit. Cost control is especially difficult when a project includes retention of existing building fabric because the condition of the work to be preserved is very difficult to assess prior to the commencement of construction. The key to the success of a project where new work is linked to old is to balance the desires of the stakeholders in a realistic manner. Both physical and financial constraints need to be taken into account early in the design process, and reserves need to be set aside as contingencies to pay for additional construction when unanticipated problems arise.

Despite the risks and complexities of combining new construction with preservation of existing architecture, great designs and successful outcomes are possible. As illustrated by the case studies in the following chapter, the myriad technical problems can be solved, stakeholder interests can be balanced, and meaningful forms can be created within this atypical architectural genre of new construction joined to old.

# Case Studies

## Case Study Selections

The case studies discussed in this chapter exhibit a clear and visible dialogue between new and old architecture. They are individual buildings with fused-together parts, where the union of new and old forms has been resolved in overall composition as well as in detail. The majority of the studies are nonresidential buildings, easily accessible from public vantage points.

Except for a few notable examples, the buildings included have not been widely published. Projects such as the addition to Frank Lloyd Wright's Guggenheim Museum in New York have been avoided, since the polemics bolstering the stakeholders' arguments were so severe. Some architects believe that what finally emerged from these bitter arguments over what was an appropriate addition to Wright's masterpiece is a compromised design. This is not to say that some of the case study designs included here were executed without some degree of controversy.

Most of the examples chosen were completed during the first decade of the twenty-first century. During this time, preservation groups were firmly established and sustainable design movements were becoming mainstream; both were advocating for the retention of existing buildings but from different viewpoints. At the same time, avant-garde computer-generated forms appeared in artistically progressive cities. When joined to old buildings, these dynamic new forms produced an amalgam of extremes in architecture that had never been seen before.

Few of the examples cited here include architectural landmarks, although the original buildings were in most cases of quality worth saving. This may explain why the architects who designed the selected projects were able to take more risks when combining new with old. In all of the studies, the

The addition to the Guggenheim Museum was controversial and highly publicized.

existing structures were cleaned and repaired, and in some instances restored. Care was taken to perform this work in ways that did not compromise the integrity of the building's fabric, even when only portions of the original structure were retained.

This chapter is organized into four categories: small interventions, major additions, repurposed buildings, and none of the above. The majority of the case studies involve major additions, where new is clearly inserted into, on top of, or alongside old. The small interventions included are lesser additions or minor alterations to larger structures. Only a few examples of repurposed buildings are illustrated since many adaptive reuse designs are interior interventions, where a visible dialogue between original and recently added features is not apparent. The last three case studies are cleverly idiosyncratic and do not fit into any particular category but do demonstrate how new design elements juxtaposed against old can pique curiosity about and enhance the appreciation of both.

Projects have been chosen that best exhibit the three degrees of contrast outlined in chapter 3: extreme, restrained, and referential. Examples that employ extreme contrast are shown as red models in the aerial photos, restrained contrast examples are shown in purple, and referential contrast examples are shown in green.

Given the variety of opinions held by informed practitioners and armchair critics concerning how new should meet old, it would be nearly impossible to find a set of case studies that would please everyone. However, the intent of this chapter is to show good work only, work that supports the thesis of this book, namely that design integrity is essential for exemplary architecture. As such, all of the case studies selected employ some degree of contrast between new and old, with clear time stamps on each respective part. Buildings that are products of more than one era receive little critical notice, but as the case studies will show, works in this genre are worthy of participation in architecture's philosophical discourse.

Extreme contrast is represented by red.

Restrained contrast is represented by purple.

Referential contrast is represented by green.

# Small Interventions

As the following case studies demonstrate, complexity and size are not prerequisites for revealing how new construction interacts with old building fabric. In fact, the small scale of the designs in this section puts into focus the purpose of the new intervention since the functional parameters that govern the alteration of the original structure are straightforward and easy to perceive.

# Dovecote Studio

Haworth Tompkins

Snape Maltings, Suffolk, England, 2009

Extreme contrast

Aerial imagery provided by Infoterra Limited.

In 1952, British composer Benjamin Britten and tenor Peter Pears founded the Aldeburgh Music Club in the pastoral countryside of Suffolk, England. The club evolved into an internationally known music center and eventually expanded into Snape Maltings—a collection of industrial buildings originally used for malting of barley. The malting operations ceased in the mid-1960s after over a century of use, and the fallow buildings fell into disrepair. Listed as Grade II structures, a category of historic building in England deemed worthy of preservation but not of the highest historic importance, the complex was repurposed beginning in 1967 as a campus for music education and performance.

The Snape Maltings grounds included a *dovecote*—typically a small structure built to house pigeons or doves, which were used traditionally for food (eggs and meat). Although the redbrick building was run down and its roof had collapsed, it was an attractive ruin and a romantic landmark for attendees of the yearly Aldeburgh music festival.

As one of the final pieces of a master plan for the campus, architects Haworth Tompkins inserted a modern structure into the aged envelope to be used as a studio for visiting artists. The new intervention is a steel *monocoque*—a single shell structure whose skin is self-supporting, much like a ship's hull. It was prefabricated off-site and then hoisted into the ruin by a small crane in one well-choreographed move. The silhouette of the design matches the outline of the original dovecote, but it is rendered in an unambiguously new material: COR-TEN steel. Over time, the steel will take on a finish compatible in color with the red brick.

The inside of the studio is open and contains a small kitchen and a sleeping loft. The interior fit-out consists of well-detailed spruce plywood panels, and the space between the new shell and the old brick is open to a ground-level drainage channel.

The dovecote had become an attractive ruin and was a romantic landmark for attendees of the yearly Aldeburgh music festival.
Courtesy of Haworth Tompkins

Like an upside-down boat, the new prefabricated monocoque was inserted into the ruin.
Photographer: Philip Vile

Following the strictest preservation principles, the ruin was treated with the utmost care. According to the architects: "Only the minimum necessary brick repairs were carried out to stabilize the existing ruin prior to the new Dovecote Studio structure being inserted. Decaying existing windows were left alone and vegetation growing over the dovecote was protected, allowing it to continue a natural process of aging and decay."[1] The patina of time was allowed to persist as the original structure accepted an addition, giving the dovecote a functional purpose once again.

The resulting balance of new and old yields a solution acceptable to stakeholders who do not often share a common viewpoint. For preservationists, the old fabric has been carefully preserved, and the new intervention is appropriately scaled and clearly differentiated from it. For modern architects, the form is crisp and the techniques employed to build it are clever. For the music patrons of Snape Maltings, the romance of the deteriorated dovecote remains—enhanced perhaps by the counterpoint of the distinctly contemporary addition.

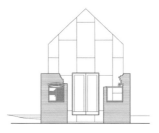
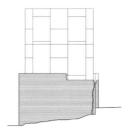

Elevation drawings
Courtesy of Haworth Tompkins

The patina of the new structure
complements the brick of the
existing form.

Photographer: Philip Vile

A corner window, with sash sliding
into the plywood interior casing,
punches through the new skin.

Photographer: Philip Vile

A new opening, clearly expressed with toothed-in brick and a concrete lintel, was cut into the ruin.

Photographer: Philip Vile

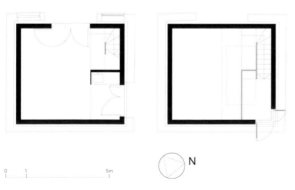

0  1                    5m

N

Plan drawings

Courtesy of Haworth Tompkins

# Hutong Bubble 32

MAD Architects

Beijing, China, 2009

Extreme contrast

Satellite imagery provided by GeoEye (www.geoeye.com) and i-cubed (www.i3.com).

*Hutongs* are narrow alleys found in the old, mostly poor districts of Beijing. They are so ubiquitous in these areas that the neighborhoods formed around them are also called hutongs. In these neighborhoods, sleeping and living quarters are packed against one another, and sanitary facilities are located in shared buildings. Many of these communities have been razed, and the residents have been displaced to generic, high-rise apartment buildings. In many Chinese cities, there has been a backlash against such demolition and the subsequent eradication of a way of life that is often cherished.

China's rapid economic growth has also had an impact on the hutongs that remain. There is concern over sanitation problems resulting from communal toilets, and in Beijing, tourists have overrun the hutongs looking for a glimpse into traditional Chinese culture and causing a disruption of everyday life.

MAD Architects, an internationally recognized Beijing firm, has devised a novel solution. Ma Yansong, a principal at MAD, grew up in a hutong and has witnessed the dramatic changes firsthand. Ma has proposed a series of "bubbles"—curvilinear, stainless steel enclosures—to be inserted among the hutong structures, providing basic needs that the individual dwellings lack. The idea was first revealed at the 2006 Venice Biennale as a vision for "Beijing 2050." Hutong Bubble 32 is the first to be built. It contains a bathroom and a staircase connecting the living space to a courtyard above.

The small, polished, bloblike form clearly contrasts with the masonry walls and clay tile roofs of the hutong. Like most avant-garde designs, Hutong Bubble 32 (at 32 Beibingmasi Hutong) has its detractors, whose arguments are reminiscent of those made by Western preservationists (e.g., lack of respect for context, contrast with existing buildings is too harsh). Others love the distinguished

Viewed from adjacent rooftops, the bubble is clearly differentiated from its surroundings.

Photographer: Shu He

A concept model of Beijing in 2050 shows a complex of stainless steel "bubbles," which serve the everyday needs of hutong residents.

Courtesy of MAD Architects

shape, and, somewhat ironically, the bubble has become a tourist attraction itself. For the client, the new structure serves her everyday needs without conflicting with the hutong lifestyle—the primary goal of the design.

In a way, the extreme contrast of the bubble's design mirrors a condition of many extremes. The bubble is a tiny, ultramodern form juxtaposed against humble vernacular architecture that is in danger of becoming extinct. It is a reaction to problems caused by extremes of modern development in a fast-growing economy. It is not an addition to a modern cultural institution—it is a simple, private bathroom that has captured the attention of the art world.

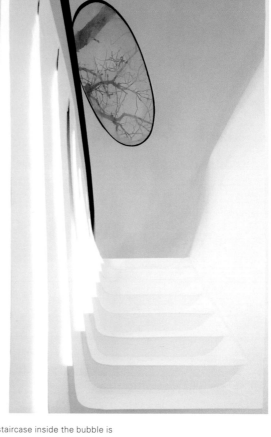

The entrance into the bubble is
a simple displacement of the
membrane's edge.
Photographer: Shu He

The staircase inside the bubble is
cast into its skin.
Photographer: Shu He

**opposite:** The bubble is a tiny
ultramodern form juxtaposed
against humble vernacular
architecture.
Photographer: Shu He

# Bar Guru Bar

KLab Architects

Athens, Greece, 2005

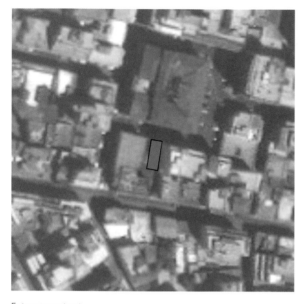

Extreme contrast

Satellite imagery provided by GeoEye (www.
geoeye.com) and i-cubed (www.i3.com).

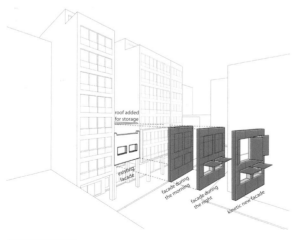

Concept diagram

Courtesy of KLab Architects

In a country in which architecture can be thousands of years old, the original lower half of Bar Guru Bar is relatively modern. It is located in one of Athens's oldest districts, Psiri. Most of the buildings date from the late-twentieth century, built during a time of urban renewal when housing for post-WWII working-class residents was simple and mundane—primarily whitewashed concrete. Psiri experienced a renaissance in anticipation of the 2004 Olympic Games and became known for avant-garde night-life. Like most edgy urban locations, the district is a mix of ordinary activity and low-level crime. KLab Architects designed Bar Guru Bar to reflect this environment.

*KLab Architecture* stands for Kinetic Lab of Architecture, a group of young architects based in Athens. Bar Guru Bar lives up to this moniker: its most visible feature is a large steel door that moves vertically. It shrouds the storefront during the day and opens for the nighttime hours. The surface is made of rusting steel—indicating a meaning beyond its functional duty. According to KLab: "The rusting material of the facade is a metaphor for the transformation in a deteriorating phase. The building is also transformational with the kinetic movement of steel plates that open to form windows and doors."[1]

Although its popularity developed mostly due to the music and food found within, Bar Guru Bar's juxtaposition of new kinetic elements and old architecture reinforces its unique qualities. Unfortunately, the bar was not sustainable despite its sudden fame in the design world. The neighborhood, true to its long history, degenerated after the bar's opening and became a dangerous nighttime location. In 2009, the giant rusted door was shut with hopes to reopen if conditions change in the future.

Openings in the giant steel door
open and close in an array of
configurations—in this case, slits
and awnings.
Courtesy of KLab Architects

# Ozuluama Penthouse

Architects Collective in collaboration with at. 103

Mexico City, Mexico, 2008

Referential contrast

Aerial photography provided by i-cubed
(www.i3.com) and Aerials Express

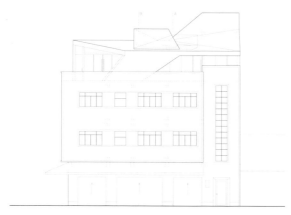

Elevation drawing

Courtesy of Architects Collective

The Hipódromo Condesa neighborhood is a formerly aristocratic enclave in the heart of Mexico City. Much of the district's architecture dates from the 1940s, tracing its roots to European modernism. The Ozuluama residence was built by an engineer in 1945 and was converted to an "artist-run space" in the 1990s, becoming a catalyst for the rebirth of culture in the district.[1]

Kurt Sattler of the Architects Collective stayed in the servant's quarters, a shack built on top of the building, during an Austro-Mexican exchange developed under the auspices of the Austrian Cultural Ministry. A decade later, he provided the design to replace the dilapidated penthouse. The new addition, in the words of his firm, "was designed to reflect the movements of its transient inhabitants in an origamilike morphology."[2]

The origami structure is created by folded plates that become both walls and roof. In the penthouse, these faceted shapes make reference to the seamless cement plaster forms of the original building below it. The uniform color of the addition recalls the monochromatic appearance of the modernist building, further reinforcing the referential relationship between new and old.

The Ozuluama penthouse is clearly modern in its use of material, its form, and its spatial definition, yet it fits agreeably on top of a distinctly different base. Rooftop architecture usually consists of a jumble of forms created for various functional reasons; when viewed from a distance, these structures tend to blend into the urban fabric. The penthouse is unique—it is a well-conceived addition, comfortably fused onto a building that connects with its ground-level surroundings, but one that distinctively stands out when viewed from afar as the nomadic form the architects intended it to be.

The penthouse appears like a
nomadic structure sitting above
the diverse topography of the city.
Courtesy of Architects Collective

The origamilike form, constructed
of Corian, stands in contrast to the
surrounding urban landscape.
Courtesy of Architects Collective

Ozuluama Penthouse

# Il Forte di Fortezza

Markus Scherer and Walter Dietl

Fortezza, Italy, 2007

Referential contrast

Satellite imagery provided by Google, GeoEye
(www.geoeye.com) and TeleAtlas (www.
teleatlas.com). (© 2009 Google, © 2010
TeleAtlas, image © 2010 GeoEye)

Completed in 1838 by the Hapsburgs, Il Forte di Fortezza is located in the southern Tirol region of northern Italy. It is an enormous stone edifice constructed to defend the Eisack Valley during the reign of the Austrian Empire. In 1918, the Italians took control of the fort, and as recently as 2003, the Italian army occupied the site. Although it was touted as one of the strongest forts in Europe, it never came under fire.

The fort consists of three massive lobes of granite construction that follow the contours of the alpine terrain; the lobes are interconnected by a maze of underground passageways. As large as a small town, the fort now serves the region as a cultural center. In 2007, architects Markus Scherer and Walter Dietl began restoration of the complex, adding elements to accommodate the center's activities.

The original architecture is heavy and powerful with a distinct medieval ambiance, despite its relatively modern origins. Granite masonry walls, vaulted ceilings, stone stairs, and rough-hewn passageways predominate. Architectural critic Catherine Slessor describes the work by Scherer and Dietl: "The thick granite walls were restored, roofs waterproofed, and windows repaired. Walled-off spaces were opened up and unsympathetic later additions removed. Throughout, the process has been a tactful cleaning up and drilling down to the raw form and structure of the fort, which itself acts as a cue for the new interventions."[1] The architects' interventions are also strong. Employing the simple materials of concrete and matte-black galvanized steel, the work complements the existing architecture.

Although the new components are purely functional in nature, they are precisely located and carefully detailed. One of the most dramatic features is a seventy-foot vertical passage—cut into the native stone—that contains stairs and an elevator to connect the subterranean caverns to the fort

above: The expansive fort complex was constructed to defend the Eisack Valley during the reign of the Austrian Empire.
Photographer: René Riller

right: A two-level, L-shaped bridge completes the circulation loop between exhibition spaces.
Photographer: René Riller

The bridge decks are channel-shaped steel gangways that are tied back to the structure with tension rods.
Photographer: René Riller

above. The shaft culminates in an entrance pavilion—a small concrete and steel building that also includes restrooms. This utilitarian structure is set against a partially deteriorated stone-wall enclosure in a manner that completes the composition both artistically and functionally.

For a circulation loop between exhibition spaces, Scherer and Dietl designed a daring, two-level, L-shaped bridge over a lake that abuts the fort at a lower level. Connecting two facades of an inside corner, the bridge is made of channel-shaped steel gangways that look as if they have been shoved out of openings in the masonry until they intersect and then are tied back to the walls with tension rods.

The interventions seem to have been made only where necessary to solve a practical problem, yet they are thoughtfully rendered in two basic modern materials so as to be differentiated from the original construction. At the same time, the new forms are almost military in their bearing, fitting comfortably into the fortress context as summarized by Slessor: "This intelligently judged reciprocity between architectures, eras, and functions is emblematic of the surprising rebirth of an extraordinary piece of nineteenth-century military history."[2]

The stair and elevator entrances
are contained in a small concrete-
and-steel building.

Photographer: René Riller

A seventy-foot vertical circulation
shaft cuts into the original stone,
connecting subterranean caverns
to the fort above.

Photographer: René Riller

The entrance pavilion combines
the new polished construction
with the existing partially
deteriorated stone-wall enclosure.

Photographer: René Riller

# Major Additions

The motives for adding a significant new form to a building vary widely, ranging from the need for more floor area to the complete reworking of the interior to accommodate a new program. A large addition to a building, however, substantially alters the visual perception of the site as the new hybrid takes on a character very different from the original. The case studies in this section illustrate how these major changes can be managed, resulting in visual balancing acts of new and old elements.

# Knocktopher Friary

ODOS Architects

Knocktopher, Ireland, 2006

Restrained contrast

Aerial imagery provided by Ordinance
Survey Ireland.

The Knocktopher Friary is a Carmelite monastery in County Kilkenny, southeast Ireland. Like all monasteries, the friary is a quiet place of contemplation. So when ODOS Architects was commissioned by the Carmelites to design a new friary—attached to an existing nineteenth-century, basilica-style church—on a site that has been a religious seat since the fourteenth century, they created a simple, traditional design.

The completed work, as noted by critics, does not explicitly reveal its composition: "Easy to overlook when skimming the images, the plan shows the masterful way in which the architects have unified friary and church with a new residential cloister."[1] The new cloister engages the church by alignment of its west wall with the centerline of the nave. ODOS then continued the elevation around the apse end of the church to create office spaces. This form connects with yet another existing building on the site: the original friary—a simple, gabled, whitewashed structure. The composition is a knitting together of two original forms with a ribbon of concrete, glass, and wood.

The unadorned simplicity of the old architecture is echoed in the new, albeit using a minimalist vocabulary. As described by critics: "Externally the palette of materials was deliberately kept to a minimum. Full height panels of treated cedar wood boards, board marked in situ concrete, and clear double glazing sit on a plinth of fair faced concrete, the top of which extends the datum of the church floor throughout the site."[2] The architects further express the church floor datum through a change of material: concrete meeting wood, concrete meeting glass, and concrete meeting concrete with slightly different finishes.

The new elevations never touch the old facades with a solid-to-solid intersection—the new is either set back from the old or the joint is glazed.

Although the new facades are comprised of wood, concrete, and glass, the level roofline of the cloister and the level datum line below define the horizontal boundaries of a carefully controlled vein of construction, especially when set against traditional, gable-roofed structures. The result is a clear differentiation between new and old, both of which are remarkably quiet architecturally—fitting for a monastery.

top: The new cloister and offices engage both the old friary and abbey church but leave the space between the existing structures open.
Photographer: Ros Kavanagh

bottom: From inside the cloister courtyard, the interaction of the new structure with the old is clearly expressed.
Photographer: Ros Kavanagh

The friary sits on a concrete
platform, maintaining the floor
level of the original church.
Photographer: Ros Kavanagh

In the parlor room, a new
unadorned wall is a backdrop
for an antique sideboard.
Photographer: Ros Kavanagh

opposite: The end elevation
shows the transparent joint
connecting the new and old
friaries.
Photographer: Ros Kavanagh

Knocktopher Friary

# Walden Studios

Jensen & Macy Architects

Geyserville, CA, 2006

Restrained contrast

Aerial photography provided by i-cubed
(www.i3.com) and Aerials Express.

Sonoma County in Northern California is known for its pastoral terrain, Mediterranean climate, and rolling hills, but it is overall less well-known than its adjacent neighbor: Napa. Ranches still exist in Sonoma, and none is more idiosyncratic than Oliver Ranch—a one-hundred-acre spread in Alexander Valley that is renowned for its sculpture and modern architecture.

Since 1985, the owners, a San Francisco Bay Area builder and his family, have commissioned site-specific pieces by some of the world's most creative contemporary sculptors.[1] "Motivated by a love of art, a desire to support artists, and also a wish to circumvent the 'business' of art," the owners, according to the ranch, were "disillusioned with valuation in the world of art collecting, and decided to commission site-specific installations that could not be moved, and therefore, neither bought nor sold. Under these conditions, the focus would be on the art itself, not its assessment."[2]

In 2001, Jensen & Macy Architects were hired to repurpose a large, one-hundred-year-old concrete barn, located in the outskirts of nearby Geyserville, into a multiuse building to support ranch activities. The program called for studios, living space, offices, and flexible outdoor/indoor areas to be used for entertaining. Architect Mark Jensen described the process: "Through a series of surgical maneuvers, we tried to open up the box and bring the landscape into it."[3]

Jensen cut into the original utilitarian mission-style structure and inserted a rectilinear glass enclosure that opens out of the old barn roof toward the rolling hill landscape beyond. The base of the existing structure was opened up as well, revealing the edges of the glass box as it threads its way through the formerly solid structure. The transparent/new element has been masterfully inserted into the solid/old structure with just enough of the original

Historic photograph of Sunsweet
Prune barn, ca. 1936
Courtesy of the Healdsburg Museum and
Historic Society

A window cutout in the
existing wall underscores the
surgical approach of the new
intervention.
Photographer: Richard Barnes

fabric remaining to demonstrate that the building is not entirely new.

Like the sculpture arrayed on the ranch a few miles away, the building sits on its site in a purposeful manner. Rather than a structure reacting to the site, however, as is the case with the sculptures, the site was massaged to react to the structure (since it was already there). Here, Jensen collaborated with San Francisco–based landscape architect Andrea Cochran, who designed the grounds immediately around the building. Taking cues from a functional requirement to protect the structure from flooding, Cochran conceived "the idea of 'piers' extending out into the 'sea' of grapes."[4] The piers form raised platforms of various materials that are distinguished from the native vegetation of the ranch. Just as the new elements of the structure clearly contrast with the old in a manner that is not overpowering, the new landscape elements clearly contrast with the untouched surroundings.

The rectilinear glass element is set back from the original concrete facade, a clear expression of new and old.

Photographer: Jeremy Jachym

The barn's roof was removed to allow light into the center of the space, but the original end wall remains visible from within.

Photographer: Richard Barnes

The one-hundred-year-old
concrete barn is punctuated
by a new transparent box.

Photographer: Richard Barnes

The new landscape is also
differentiated from the
surrounding vegetation.

Photographer: Marion Brenner

# Contemporary Jewish Museum

Studio Daniel Libeskind

San Francisco, CA, 2008

Extreme contrast

Aerial photography provided by i-cubed
(www.i3.com) and Aerials Express.

Daniel Libeskind has designed a number of Jewish museums outside of the United States and brings to each his easily recognizable signature. Given that his portfolio includes many other cultural institutions and that his work has engaged many historic buildings, he was a logical choice for San Francisco's Contemporary Jewish Museum. Since his work is known for acute angular forms, it is not surprising that he employed his distinctive virtuosity in the design for this project.

The building he transformed was originally constructed in 1881 as a utility substation for Pacific Gas and Electric Company (PG&E). A few years after the 1906 earthquake, it was remodeled by Willis Polk, who ran the local office of D. H. Burnham & Company and became one of the most prolific architects of postearthquake San Francisco. The building is a large, simple masonry structure with highly ornamented entrances—a motif Polk employed in other PG&E buildings as well. Although Polk's designs are typically less reserved, the building was never intended to be more than a humble, industrial structure. The substation is a recognized landmark and on the National Register of Historic Places.

Libeskind inserted two stalactitelike forms into the existing structure. One protrudes above the roof and is not immediately perceived from the entrance courtyard. The second extends horizontally out of the south end of the building into a smaller, subordinate courtyard (part of a pedestrian pathway connecting the city's retail and financial districts to a large civic plaza in the convention center district). The crisp, angular shapes are clad in iridescent-blue stainless steel panels and are meticulously detailed, including the joints between the new metal skin and the old masonry one. Formally, the disparity between new and old shapes could not be more pronounced, and the choice of precise metal juxtaposed against rough masonry reinforces the idea of contrast.

The two stalactitelike forms contrast greatly with the urban surroundings.
Photographer: Bruce Damonte
Courtesy of the Contemporary Jewish Museum, San Francisco

San Francisco is truly a liberal city, well known for tolerance at many levels. Paradoxically, however, it is very conservative architecturally, and it is surprising that the museum's extreme contrast between new and old was allowed by city agencies. Although the scale of the new form protruding from the masonry box was diminished during the approval process, what survived remains a strong formal statement.

Not everyone is convinced that the juxtaposition results in a successful design. Edward Rothstein, writing for the *New York Times*, expressed: "Its skewed geometries are unsettling; the effect is more vertiginous than harmonious."[1] Harmony, however, may not have been the design intent; in fact, the aim could have been to create an abstract definition of space, which is common to Libeskind's work. To this end, the design is successful. As a contribution to the urban design of the area, the building is unarguably effective. One of a number of cultural institutions in a redeveloped part of San Francisco, the museum has unequivocally enhanced its site, especially on the south end. The light plays off of the new form into a space defined by a number of undistinguished buildings, giving meaning to a pause along a pedestrian pathway that connects important parts of the city.

The relationship between new and old architecture is crystal clear in this building. Libeskind's signature is easily recognizable here, and it has been executed with a high degree of craft—just like the original substation with Polk's easily recognizable signature—and the resulting combination adds a meaningful urban space to the city.

The angular form protruding from the south end of the building is much more pronounced than the original roof form.

The additions contain openings for doors, windows, and skylights; only the doors are orthogonal.

**opposite:** At night, the forms play to the lights of the city.

Photographer: Bruce Damonte
Courtesy of the Contemporary Jewish Museum, San Francisco

# Morgan Library

Renzo Piano Building Workshop

New York, NY, 2006

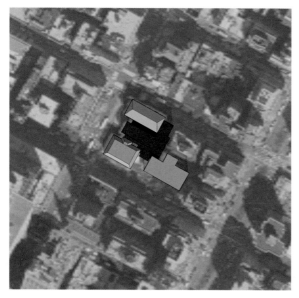

Restrained contrast

Aerial photography provided by i-cubed
(www.i3.com) and Aerials Express.

Renzo Piano is an internationally recognized architect, known for his reserved, understated modern style and unwavering attention to detail. Both are evident in his addition to the Morgan Library in New York, where he inserted a new element to knit together a jumble of three historic buildings: the 1852 mansion by Isaac N. Phelps Stokes, the 1906 library by McKim, Mead, and White, and the 1928 annex by Benjamin Wistar Morris.[1] The buildings are distinct in appearance, representative of the different times in which they were constructed. The Italianate brownstone mansion is typical of its era as is the original, but contrasting, Beaux Arts library. The annex is stylistically similar to the library because its design was based on a plan conceived by Charles McKim.

Piano met the challenge of the site by creating a central atrium in the courtyard shared by the historic buildings plus three subordinate elements: a connector between the 1906 library and its annex, an entrance pavilion, and a four-story administrative support building fronting 37th Street. He viewed the courtyard space as a kind of piazza, a theme he has previously employed to organize museum plans. In this case, the piazza is an enclosed element, rendered in a modern vocabulary in contrast with its surrounding buildings. In the words of *New York Times* architectural critic Nicolai Ouroussoff,

> The layout sets up a mesmerizing rhythm between new and old. The boxy pavilions are joined to the more massive stone buildings by vertical slots of glass. By creating a slight separation between each of the buildings, Mr. Piano allows pedestrians a glimpse deep into the central court from side streets to the north and south. It's as if the Morgan complex has been gently pulled apart to let life flow through the interiors, hinting at the fragile balance between

A simple, opaque box is inserted between two landmark buildings.
Photographer: Michel Denancé

Glass is only used where new meets old or to provide light into spaces that do not house the collection.
Photographer: Michel Denancé

the city's chaotic energy and the scholar's interior life.[2]

This textbook solution, transparent connective tissue between disparate parts, has been gracefully implemented. Since most of the Morgan's collection of rare manuscripts eschews light, bathing the central court in it defines a beacon from which to circulate among the three existing buildings. The boundaries between old and new are therefore crystal clear.

The exterior composition works in a similar manner. Rather than leaving a space between two of the landmark buildings on Madison Avenue, Piano inserted a simple, opaque box between them (containing a large reading room). The box is supported visually by slender columns, marking the entrance to the library complex. This new element contrasts with the historic buildings on each side, but at the same time, the scale of the new piece joins the adjacent buildings together as wings of a single institution. The entrance sequence completes the story: entering beneath the box sets up the light-filled experience that awaits visitors once inside the naturally illuminated central court.

Piano's intervention is a work of architectural rigor and discipline. The buildings linked by the addition are not all from the same time or architectural family, yet the contrast between the new work and the collection of older structures creates a strong connection. A new structure that attempted to blend in with the landmark buildings or one that intended to play a subordinate role would not have yielded the Morgan Library masterpiece that Piano's addition precipitated.

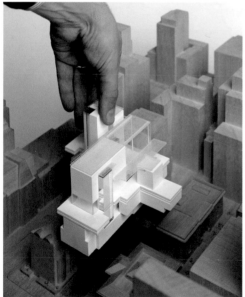

Site models

Photographer: Stefano Goldberg

Courtesy of the Renzo Piano Building
Workshop

The light-filled interior courtyard
connects the three existing
buildings.

At night, the simple modern cube
marking the entrance serves as
counterpoint to the old buildings
that flank it.

Photographer: Michel Denancé

# Moderna Museet Malmö

## Tham & Videgård Arkitekter

Malmö, Sweden, 2009

Extreme contrast

Satellite imagery provided by GeoEye (www.
geoeye.com) and i-cubed (www.i3.com).

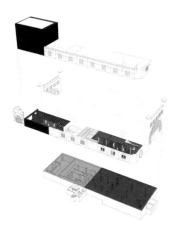

Exploded diagram

Courtesy of Tham & Videgård Arkitekter

Malmö is the third-largest, and most diverse, city in Sweden. Located in the country's southernmost province, it is separated by only a few miles from Copenhagen, Denmark. It is also an old city, with origins dating from the twelfth century. Architecturally progressive, Malmö features a blend of very old and modern buildings. The city's most famous recent structure is Santiago Calatrava's Turning Torso, a residential skyscraper and Sweden's tallest building.

The site for the Moderna Museet is a former electricity plant, built around 1900 and closed in the 1990s. Despite the building's utilitarian function, its public facades are elaborate and romantic in design, and the original structure presented a dignified demeanor to the surrounding environment. Upon closing, it became an art gallery, which served as a catalyst for an area that is now a cultural district of sorts.

The Moderna Museet is Sweden's national museum of contemporary art, based in Stockholm. As a satellite of the parent institution, the museum in Malmö was required to meet the highest infrastructure standards for display of modern art. Given a short time frame to design and build the museum, architects Bolle Tham and Martin Videgård were charged with a radical reconstruction—to create something new within the existing shell.[1]

Tham and Videgård's solution was a total makeover of the building's interior and the insertion of a dramatic entrance hall that includes a bookstore and café. The entrance hall is a distinct element, bright orange both inside and out. The internal galleries, however, are simple, white rooms—functionally appropriate for spaces that display art. On the outside, the extreme contrast between the brightly colored facade and the original industrial walls allows the new hall to be an exclamation point in the prosaic architectural context nearby.

top: The bright orange entrance hall is a striking departure from the elaborate and romantic facades of the former electricity plant.
Photographer: Åke E:son Lindman

bottom: The saturated color extends into the café, uniting interior and exterior.
Photographer: Åke E:son Lindman

According to the architects, "The colour also connects to the industrial character of the building, that kind of paint you associate with tools and vehicles, or used to protect steel from rusting.... We looked for a colour that would relate to the red-orange-brick architecture but at the same time be more abstract."[2] One could argue that hue, saturation, and brightness could have been toned down and that most passersby probably fail to see the connection. However, one could also argue that the extreme contrast between the new entrance element and the original plant facades causes them to enhance one another. Tham and Videgård's other work is full of saturated color, and in the Moderna Museet, skillful use of color has provided a brilliant architectural narrative.

opposite: The simple, direct shape and skillful use of color provides a compelling contrast to the existing entrance.

Photographer: Åke E:son Lindman

At night, the bright orange interior glows.

Photographer: Åke E:son Lindman

Elevation drawing

Courtesy of Tham & Videgård Arkitekter

# CaixaForum Madrid

Herzog & de Meuron

Madrid, Spain, 2008

Extreme contrast

Swiss firm Herzog & de Meuron is preeminent on the world stage of museum design. Few other architects would have the audacity to suggest levitating a brick mass and then piling on more visual weight, yet this is exactly what the architects pulled off at the CaixaForum museum in Madrid. The gravity-defying structure is breathtaking, and in an odd way, makes perfect sense.

The museum is housed in a former power plant, originally built in 1899. The structure was listed as a historic building of minor importance by the local heritage commission because of its role in the establishment of electrical power in Madrid. The museum's plaza occupies the site of a former gasoline station. These two structures were only a few blocks away from Paseo del Prado and the world-famous Museo Nacional del Prado, and had become misfits in Madrid's famed cultural district.

Only the shell of the original power plant was a protected historic resource, so the architects were able to remove the parts of the building that were no longer relevant—including the structure's base.[1] Lifting the brick mass of the building above the sloping ground plane created a space that is an extension of the new plaza (albeit tension-filled due to its low head-height). Removal of the base also allowed entrances and exits to be located according to the circulation pattern of the surrounding area rather than conforming to the power plant's elevations. The entire mass is supported on interior columns set back from the facade, resulting in a daring visual challenge—the glass base, also set back from the facade, accentuates this tension. Little of the original building remains. The architects were permitted to commit the ultimate preservation sin, often referred to as "facade-ism," in which only the exterior skin of the original is retained, as a trade-off for the civic benefits of a museum compared to an abandoned power plant and a gasoline station.

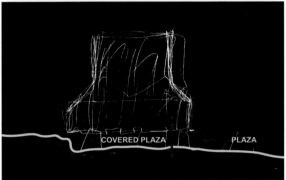

The building includes two underground levels and two new floors above the original brick shell, significantly increasing the size of the structure. The addition takes its sculptural cues from the surrounding urban roofscape, and attaches to the original roof undulations like a snug-fitting cap. Clad in rusting steel plates with a pattern incised on some walls—a technique used by the architects in other designs—the addition employs a restrained use of ornament that relates to the degree of ornamentation found on the facade of the power plant (understated when compared to other late-nineteenth-century buildings in Madrid). The color and texture of the two eras of construction are also similar, but the way the new cap rests on the outer edge of the old cornice makes it impossible to miss the joint where new and old are fused together.

Especially when viewed from the new plaza, the extreme contrast between new and old architecture improves both. The original building, lifted off of the ground and fitted with a giant hat, beckons the eye. The addition, despite its sculptural qualities, would not be as interesting without the hovering base. A third element, a sixty-foot-high vertical garden by botanist Patrick Blanc, flanks one side of the plaza and completes the composition. The juxtaposition of new, old, and living architecture of this scale exists nowhere else in the world.

top: Site before renovation
Courtesy of Herzog & de Meuron

middle: Sketch diagram
Image by Herzog & de Meuron

bottom: An extension of the plaza, the space under the lifted building features a faceted ceiling.
Photographer: Daniel Lobo for Daquella Manera

A new space is created,
extending the plaza below the
mass of the building.
Photographer: Duccio Malagamba
Courtesy of Herzog & de Meuron

The new structure locks into the
existing roofline, differentiated by
color and material.
Photographer: Duccio Malagamba
Courtesy of Herzog & de Meuron

opposite: The entire structure,
the original building and the new
addition, is lifted.
Photographer: Duccio Malagamba
Courtesy of Herzog & de Meuron

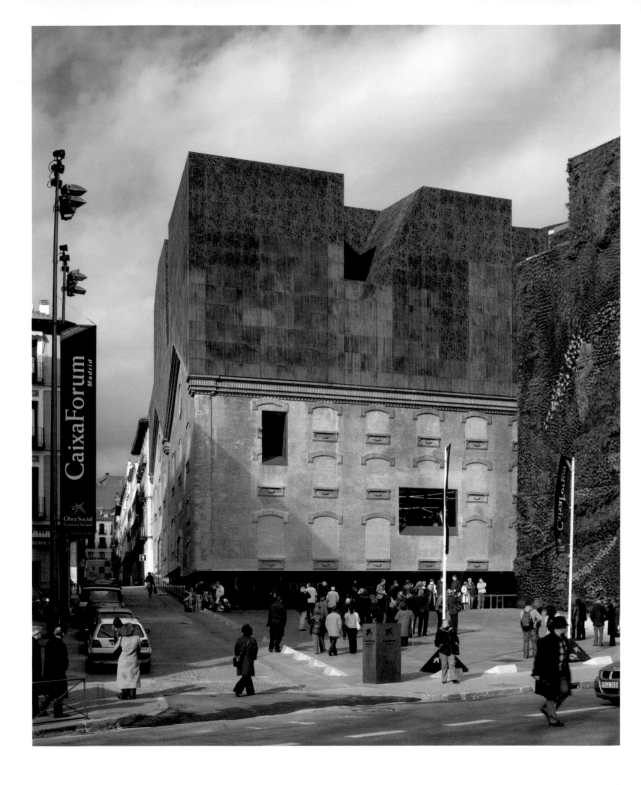

CaixaForum Madrid

# 1 Kearny Street

Office of Charles Bloszies

San Francisco, CA, 2009

Referential contrast

Aerial photography provided by i-cubed

(www.i3.com) and Aerials Express.

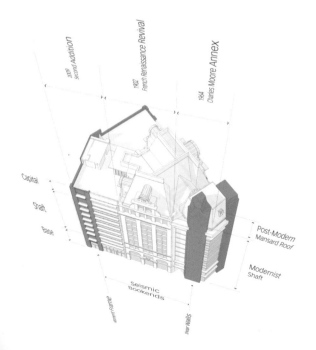

1 Kearny is a single building comprised of three fused-together pieces constructed at different times. Built in 1902 as the Mutual Savings Bank, the central piece is an opulent twelve-story building fashioned in the French Renaissance Revival style. The second is a modern twelve-story annex constructed in the mid 1960s, designed by renowned architect Charles Moore. The third piece is a new ten-story addition, designed by the author, and completed in 2009.

The owners were looking for a new building to complete a "triptych" that would enhance the distinctive character of the two existing structures. Planners were looking for a building that would increase the density of the block by filling in its "missing tooth" and wanted it rendered in a modern vocabulary that would conform to the design guidelines for the city's downtown building conservation district. Historic preservation advocates were looking for a design sympathetic to the styles of both the original building and the Charles Moore annex. The goal of the project was to satisfy all of these stakeholders.

A number of technical deficiencies needed correction as well. The 1960s structure was designed to support the 1902 building in a "served-servant" relationship, following architectural theories of the time. The building's core components, including elevators, fire stairs, and restrooms, had been located in the annex. The servant, however, was ill-positioned for the area it was intended to serve. The annex turned its back on views down San Francisco's famed Market Street, and the elevator lobby was too large for a full-floor tenant. Also, typical of many turn-of-the-nineteenth-century buildings, the original structure did not meet modern seismic requirements. Even though the annex was designed as a stiff concrete structure, it was not strong enough to brace the 1902 steel-framed building.

Historic photograph of original building, ca. 1910

Courtesy of the San Francisco History Center, San Francisco Public Library

The first addition was a corner annex designed by Charles Moore in the 1960s.

Photographer: Matthew Millman

**opposite:** The original building is seismically braced by the steel structure of the new addition and the concrete structure of the annex. Seismic diagram

These deficiencies were addressed with the new structure. The circulation core was relocated from the 1960s annex to the addition, liberating the corner with the prime view to be used as future office space. And the steel structure of the addition in concert with the concrete structure of the annex form seismic "bookends" that brace the original building without tearing into it.

The facades of the new building were designed to respect those built in the two previous eras, while also standing on their own. As local architectural critic John King observed, "Each piece is serious architecture, shaped by thoughts of how best to fit within the city in lasting ways."[1] The composition of the Market Street facade is classical, with a clearly defined base, shaft, and capital, like its immediate neighbor. Striking vertical bands of terra-cotta echo the stair tower masonry of the annex. The new facades are textured with crisp metal fins that project out as sunscreens above windows on the south-facing elevation.

The completed structure also engages the city in two different ways, both satisfying agency requirements. The roof terrace of the addition is a privately owned public open space (POPOS)—an elevated park with a breathtaking view. The interior of the main lobby was designed by avant-garde architecture firm Iwamoto-Scott, specifically commissioned as a work of public art. Known as *Lightfold*, the lobby features an undulating geometry, distinct from the rectilinear lines of the addition. It is yet a fourth contrasting element to the overall composition of 1 Kearny.

The 2009 structure is clearly differentiated from the original French Renaissance Revival building as well as the modernist annex, but it takes design cues from both, just as Moore's annex took cues from the 1902 building. The addition is carefully crafted to stand on its own quietly while making reference to both older structures.

The new seismic frames are clad in a terra cotta rain screen, an archaic material rendered in a modern vocabulary.

Photographer: Matthew Millman

top: The functional ornament of the metal sunscreens gives the new facade a texture similar to that of the original building.

bottom: *Lightfold*, Iwamoto-Scott's design for the lobby, is another evolution of the 1 Kearny complex.

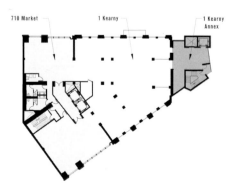

Plan drawings, before and after

opposite: The facade of the new addition references both older structures while maintaining its own character.

Photographer: Matthew Millman

1 Kearny Street

# Hearst Tower

Foster + Partners

New York, NY, 2006

Extreme contrast

Aerial photography provided by i-cubed
(www.i3.com) and Aerials Express.

Norman Foster is familiar with how new and old architectural styles interact, given his extensive portfolio of large additions to landmark buildings. The Hearst Tower completes a design commenced over seventy years ago, but in a far different manner from what was originally conceived. It is currently the largest vertical addition in the world. Foster has played on the world architectural stage for a long time—long enough to propose an audacious form that is in striking contrast with the existing structure.

In the late 1920s, William Randolph Hearst retained architect Joseph Urban, well known for his theater set creations, to design Hearst's New York headquarters. Urban designed a twenty-story building in the art deco style, rising from a dramatic, six-story podium. His art deco plan, with sculptural features wrought against angular shapes, was appropriate to capture the spirit of the decade. Due to the onset of the Great Depression, however, only the six-story base of the building was constructed.

Taking this history into account, Foster saw an opportunity to complete an unfinished design in a theatrical manner with angular facets, somewhat reminiscent of art deco flourishes. Critics, however, do not see the new design as a harmonious amalgam of new and old; rather, in the words of *New York Times* architectural critic Nicolai Ouroussoff, "Past and present don't fit seamlessly together here; they collide with ferocious energy."[1] For some, a harmonious, seamless fit is not necessarily a prerequisite for a clearly delineated hybrid design.

The tower is now forty-two stories tall, including the original six-story base. The podium structure was gutted, with only the facade and one structural bay behind it remaining. The link between new and old at the sixth floor is a horizontal ring of glazing that clearly separates the two eras of construction. The tower rises from within, supported on an

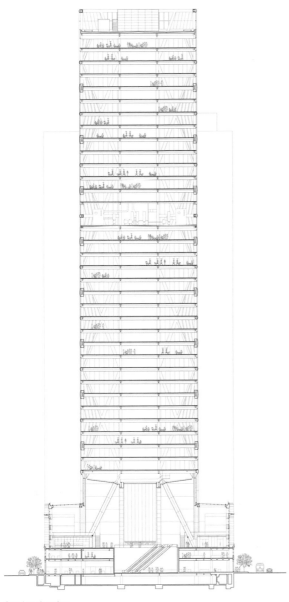

Section drawing
Courtesy of Foster + Partners

entirely new steel frame—an efficient triangulated structure that resulted in a 20 percent reduction of steel tonnage used, compared to conventional skyscraper framing. In fact, the entire building is energy efficient and is the first major building in the United States to achieve a LEED Gold rating.

Although the project had many supporters during its design phases, the scheme was not approved without controversy. Simeon Bankoff, executive director of the Historic Districts Council, asserted, "We feel it's not an appropriate building. It does not respond to, respect, or even speak to its landmark base."[2] Ouroussoff countered, stating, "Lord Foster's approach to history is frank and direct. It's as if the facades of the original building are really just there to keep out the rain."[3] Interestingly, they are both using the same point to argue opposite sides of the coin.

Strictly speaking, Foster's design approach runs contrary to accepted historic preservation precepts. The new tower overshadows the existing structure, and only the facade remains of the base. Nonetheless, completion of an unfinished work in a bold manner that harkens back to the theatrical roots of the original is compelling. Further, the new tower is not a lone individual building: it is one of a series of similarly scaled structures along 57th Street. The Hearst Corporation needed more space, and what could be more fitting than to inhabit the footprint of its original headquarters?

Hearst Tower is an example of architectural virtuosity at a scale that only an internationally famous designer could execute. New and old are clearly delineated with easily recognizable time stamps. The inherent qualities of both designs, and the extreme contrast of old and new, make this high-impact architecture.

Hearst Tower

The steel frame of the new tower
is a prominent architectural
feature in the lobby of the
existing structure; skylights
provide an impressive view of
the new through the old.
Photographer: Chuck Choi
Courtesy of Foster + Partners

opposite: The original, period
1920s structure acts as the base
for the large, modern addition.
Photographer: Nigel Young
Courtesy of Foster + Partners

Old Buildings, New Designs

Hearst Tower

# Repurposed Buildings

Oftentimes a repurposed building retains its original exterior appearance but its interior is entirely refitted with no visible evidence of the former use. The case studies included here are different—new interior elements have been inserted into the original envelope in a manner that creates a dialogue between old and new. The clarity of this dialogue is rooted in contrast, and the projects illustrated employ differing degrees of contrast to make this point.

# Village Street Live-Work

Santos Prescott & Associates

Somerville, MA, 2006

Restrained contrast

Aerial photography provided by i-cubed
(www.i3.com) and the USGS.

Adèle Naudé Santos is a highly acclaimed architect and educator with a habit of converting dilapidated buildings into her own live-work studios as she moves around the country. According to Santos, when she relocated to the East Coast, she sought out "an industrial building, preferably in awful condition, one that required some great imagination."[1] She found a run-down structure with a long-neglected courtyard, a specimen very similar to the San Francisco building she had repurposed ten years earlier. The resurrected San Francisco structure now houses her West Coast office, Santos Prescott & Associates, while the Village Street studio in Somerville, just outside of Boston, is her home and office workplace as she serves as dean of the School of Architecture and Planning at Massachusetts Institute of Technology. As an educator, Santos has made an imprint on generations of students with an emphasis on regional planning and large-scale issues. As her own architect, she is very skillful at the small scale, creating rooms that interact with quasi-outdoor spaces, as well as interiors filled with artifacts from her roots in South Africa.

The building was originally a bronze foundry, built in 1860. It was a single-story industrial structure, rectangular in plan, with brick masonry walls; it measured fifty feet by one-hundred-fifty feet. The northern third of the building was an enclosed casting room with clerestory windows and a large gantry crane supported by two steel girders that ran the entire length of the room. According to Bruce Prescott, principal in her San Francisco firm, "the goal was to retain the openness of the industrial space, the richness of the weathered brick, and the sculptured quality of the existing machinery while also making a comfortable home."[2]

The casting room was converted into a residence by adding floors between the crane rails and the exterior envelope and retaining the volume of

top: An interior courtyard, enclosed on two sides by the original brick masonry walls, separates the residence from the studio.
Photographer: Eric Oxendorf

bottom: The main living area, formerly a bronze casting room, retains the volume of the original space.
Photographer: Eric Oxendorf

the center of the space. The crane was left in place, and although dominant as a relic, it seems to fit among the colorful artifacts of the interior decor. Partitions and cabinetry are straightforward, simple designs that quietly complement the original brick walls and rough timber roof. A new radiant-heated concrete floor completes the interior composition. New and old elements are clearly differentiated but at the same time work together as a collection of warmly colored components.

The remaining two-thirds of the plan was divided into two spaces. At the south end of the site, a new studio was inserted into the brick shell by adding a roof and a glass wall, which looks toward the residential quarters. The space between the studio and residence was landscaped with fruit trees and shrubs and transformed into a garden courtyard, enclosed by the remaining brick masonry walls. A glass lean-to greenhouse structure was added to the former casting room wall to form a kind of sunroom between the garden and the residence. This transition space also acts as a thermal collector, obviating the need to insulate the brick wall.

The same three components (live, work, and enclosed garden) form the core of Santos's San Francisco complex. Both projects illustrate how old buildings, replete with architectural character, can be reclaimed to become vibrant, useful pieces of dense urban fabric once again.

The glazed facade of the studio
faces the courtyard.

Photographer: Eric Oxendorf

The separate studio building
provides ample space to work
away from the house.

Photographer: Eric Oxendorf

opposite: The gantry crane
complements the colorful artifacts
of the interior decor.

Photographer: Eric Oxendorf

Village Street Live-Work

# Selexyz Dominicanen Bookshop

Merkx + Girod Architects

Maastricht, Netherlands, 2006

Extreme contrast

Satellite imagery provided by GeoEye (www.
geoeye.com) and i-cubed (www.i3.com).

In many parts of the world, religious congregations are shrinking, leaving behind vacant architectural treasures. Built for ecclesiastical use and bound by a long architectural tradition, empty churches cannot be easily adapted for new uses. Most old churches are slender and lofty and uninsulated, listed on historic registers, and suffer from years of deferred maintenance, making them politically difficult and costly to renovate. In most cases, these beautiful and historic buildings should be saved, but doing so requires some ingenious thinking.

A Dominican church in Maastricht, dating from the thirteenth century, had been used as a warehouse before Merkx + Girod Architects were asked to convert it into a bookstore for Selexyz, a large Dutch bookseller. Since the footprint of the church could not accommodate the bookstore's inventory, the architects proposed a vertical design—a staircase within the shelves allows browsers to peruse books much as they would in a single-level store. A number of benefits resulted from the vertical orientation, including the ability to view the expansive nave from perspectives previously unavailable and closer visual access to the historic paintings on the vaulted ceiling. The architects carefully placed the interior fittings within, but not attached to, the pre-Gothic interior. The new elements are straightforward, simple, functional, and clearly differentiated from the aged construction.

The conversion of the church into a bookshop is successful on many levels. The repurposed building continues to serve the community as a source of wisdom as it has for centuries, the edifice has retained its historic identity, the book stacks fit into the nave remarkably well, and the artistic features located in hard-to-see places are more easily viewed.

A towering, three-story black steel bookstack was installed in the long, high nave.

Photographer: Bart Gerritsen

Selexyz Dominicanen Bookshop

# California College of the Arts

Leddy Maytum Stacy Architects (1999)

KMD Architects, Jensen & Macy Architects,

David Meckel (1996)

San Francisco, CA

Restrained contrast

Satellite imagery provided by GeoEye (www.
geoeye.com) and i-cubed (www.i3.com).

Founded in 1907, California College of the Arts (CCA) has produced a number of renowned artists and has evolved into a leader in education for architecture, fine arts, design, and writing. In 1987, the college expanded into San Francisco from its main campus in Oakland. Due to the success and growth of the San Francisco–based design programs, CCA adapted an industrial building complex to accommodate a growing student population in the early 1990s.

The first phase of expansion included repurposing a large, concrete building, originally designed by Skidmore, Owings, and Merrill (SOM) in 1950 for the Greyhound bus company. The building had served as a Greyhound maintenance facility but had been vacated after sustaining damage during the 1989 Loma Prieta Earthquake. The structure consists of two pieces—a two-story building, rectangular in plan with an industrial-style saw-toothed roof, and a large, open, maintenance shed with a roof supported by post-tensioned concrete arches covering a space the size of a football field.

David Meckel, founding dean of the architecture program at CCA, assembled a team of architecture faculty to convert the two-story segment into design studios, classrooms, an exhibit gallery, a library, and administrative offices. The rectangular plan with ribbon windows easily accommodated the program, leaving seismic strengthening and repair of the structure as the primary challenges.

With the author acting as structural engineer, in concert with interior architects Jensen & Macy, a system of seismic braces was devised for the core of the structure, separating the various spaces without interfering with the perimeter light. A layer of movable partitions and lighting modules completed the interior. The steel braces and industrial-style fittings work together to define the raw space without compromising the industrial muscle of SOM's original design.

top: Historic photograph, ca. 1990
Photographer: Richard Barnes
Courtesy of Leddy Maytum Stacy Architects

bottom: The college inhabits
the industrial building originally
designed for the Greyhound bus
company by Skidmore, Owings,
and Merrill in 1950.
Photographer: Richard Barnes
Courtesy of Leddy Maytum Stacy Architects

The second phase of the project commenced a few years later. Leddy Maytum Stacy Architects (LMS) was selected to transform the vast maintenance shed into the centerpiece of the San Francisco campus. Here again, seismic strengthening was a challenge since the huge concrete arches defined an interior envelope of uninterrupted space. Working with the local office of world-renowned engineers Arup, LMS inserted tubular braces that subdivide the expanse without altering its powerful impact. New studio areas, flexible classrooms (with hinged partitions), a café, and an exhibit hall were introduced in a similar manner. The central spine was left open and now functions as a large critique space known as the "nave."

As a leader in sustainable design, LMS employed the large roof structure as a source for solar power. Hot-water solar collectors supply energy for a radiant floor, added as a concrete layer over the industrial surface formerly used to maintain Greyhound's fleet. The building has been acknowledged as one of the most sustainable retrofit projects in the country.

No expansion of the exterior envelope was made, but the juxtaposition of new and old design is evident nonetheless. (The old in this case is not so old, but it is an exemplary work of industrial modernism.) All of the insertions and interventions are clearly differentiated from the existing. The combination of precast and poured-in-place concrete, along with large expanses of glass, honestly expresses the building's original purpose. The steel bracing elements, as well as the contrasting interior components, reveal the story of the building's transformation. It is a fitting home for a progressive art school.

Steel braces were inserted into the concrete envelope creating studio spaces and serving as earthquake resisting elements.

Photographer: Richard Barnes

Precast concrete arches and steel braces reveal how new construction meets old.

Photographer: Richard Barnes

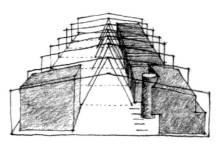

Concept sketch
Courtesy of Leddy Maytum Stacy Architects

Studio spaces and seismic braces
line the enormous shed structure.
Photographer: Karl Petzke
Courtesy of California College of the Arts

# None of the Above

The final three case studies share one design attribute—clarity—but otherwise they are unlike each other and do not fit into any of the previously defined three categories. Each is cleverly conceived and executed to reveal what is old and what is new in three distinct ways.

# 185 Post Street

Brand + Allen Architects

San Francisco, CA, 2008

Restrained contrast

Aerial photography provided by i-cubed
(www.i3.com) and Aerials Express.

Historic photograph, 1918
Courtesy of San Francisco History Center,
San Francisco Public Library

"Modernized" 1950s remodel,
ca. 1990s
Photographer unknown
Courtesy of Brand + Allen Architects

**opposite:** The glass-clad building
anchors a prominent retail corner.
Photographer: Mariko Reed
Courtesy of Brand + Allen Architects

A building does not typically shed its skin, but the structure at 185 Post Street in San Francisco has done so more than once. One of many structures in the downtown core built just after the famous 1906 earthquake, it was constructed of brick masonry as a reaction to the devastating fire that followed the seismic shock. In the late 1950s, the building's original brick facade was overclad with a tile-and-metal-framed window system in an attempt to modernize the street elevations.

185 Post is located on a prominent corner in a building conservation district, where strict controls govern designs for alterations of existing structures. Brand + Allen Architects proposed a clever solution that took into account the disparate desires of the typical stakeholders. They removed the 1950s skin to expose the original brick facade. It was heavily scarred from the previous tile installation, but presented a simple, unadorned surface with large punched openings. The architects then wrapped this masonry envelope in glass. The result is a beautiful reflection of past and present. As architectural critic John King observed, "New glass walls encase a six-story masonry building from 1908...without window frames or mullions, so the effect is that of a ninety-five-foot-high display case pulled tight across the past."[1] The ground-floor retailer, a diamond jeweler, is an ideal fit for a building that is a jewel case itself.

The face of the historic building was preserved, albeit stripped of its original 1908 character. A sleek, modern building has been added to a collection of historic structures without disrupting the overall makeup of the urban scene. Preservationists, urban planners, and modern architects are equally pleased in a rare convergence of opinion.

185 Post Street

# Hôtel Fouquet's Barrière

Édouard François

Paris, France, 2006

Extreme contrast

Satellite imagery provided by GeoEye (www.
geoeye.com) and i-cubed (www.i3.com).

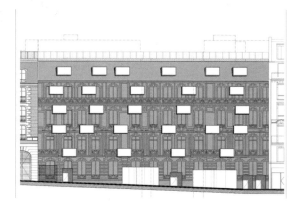

Elevation drawing
Courtesy of Édouard François

**opposite:** Rendered in modern
materials, the addition to the
hotel is an exact copy of a typical
Haussmann facade.
Courtesy of Édouard François

Édouard François has designed an addition to a venerable Parisian hotel in a most unusual manner. The Hôtel Fouquet's Barrière raises a thought-provoking question concerning the integrity of an architectural design where new meets old.

One reason Paris is an architectural delight is the unrelenting imprint of Baron Haussmann. Beginning in the mid-nineteenth century, Haussmann divided Paris into the *arrondissements*, or districts, modernizing the medieval city through works of civil engineering—including Paris's famous sewers. Haussmann's most lasting legacy, however, was a uniformity of architecture—perhaps best embodied by the ubiquitous five-story walk-up apartment building.

Charged with the renovation of Hôtel Fouquet's Barrière, François proposed a novel idea he called "copy + edit." His addition to the hotel is an exact copy of a typical Haussmann facade but created in modern materials. At a quick glance, the structure appears to have been built in the 1880s, but upon closer inspection, it becomes clear that the facade is of contemporary concrete construction. Its ornamental details are precise replicas of original decorative features. The "edit" part of François's concept consists of inserting ultramodern windows into the pseudohistoric facade. The result is a jarring juxtaposition of seemingly old architecture and openings randomly cut into the facade that in certain light conditions resemble flat-screen LED displays.

Is this a serious work of architecture or is it a folly? Most agree that it is a provocative design that has been skillfully executed. Modern incisions into a historic facade would leave a clear indication of what is old and what is new, but what about modern incisions into a modern facade rendered in a historic motif? Has design integrity been replaced by an avant-garde gesture?

Hôtel Fouquet's Barrière

# Recycled Batteries

Office of Charles Bloszies

Pacific Coast, Northern California (unbuilt)

**Extreme contrast**

Satellite imagery provided by Google and Terrametrics (www.truearth.com). (© 2010 Google, image © 2010 Terrametrics).

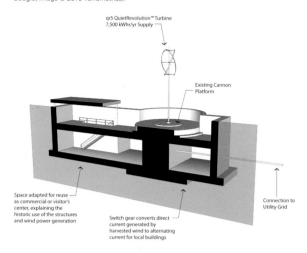

qr5 QuietRevolution™ Turbine
7,500 kWhr/yr Supply

Existing Cannon Platform

Space adapted for reuse as commercial or visitor's center, explaining the historic use of the structures and wind power generation

Switch gear converts direct current generated by harvested wind to alternating current for local buildings

Connection to Utility Grid

Diagram of wind turbine integrated with battery

**opposite:** The historic batteries provide the structural framework, and the ideal oceanfront location, for the wind turbines. Concept rendering

In some instances, old structures may be able to be saved by new, sustainable interventions. The historic concrete bunkers along the Pacific Coast are an example. The bunkers, built by the U.S. military from pre–Civil War until the late twentieth century, were provisioned for use but never called into action. Many of these historic structures, also known as batteries, have fallen into disrepair due to their proximity to the ocean and exposure to salt air. But because they are situated with a clear trajectory to the sea, the batteries enjoy steady wind exposure, ideally suited for generation of wind power.

The author has proposed erecting vertical wind turbines on the batteries near the Presidio of San Francisco. Currently in the United States, wind farms are typically constructed far from the point of use for the energy generated, but the small-scale approach proposed for the batteries would generate power very close to where it would be consumed. The now-defunct cannon bases could serve as support for the turbine towers, and the interiors of the batteries could house the switch gear needed for transforming the wind-generated direct current to alternating current, which could then be used to power nearby buildings. This proposal is a kind of swords-to-plowshares approach for adaptive use of a historic structure that has little hope of being saved by other means.

An idiosyncratic suggestion such as this one may be more symbolic than practical, but it serves to illustrate how new and old can be melded together in a manner that is mutually beneficial. The turbine receives a foundation at no energy cost, and the battery itself is recharged with a new function.

Recycled Batteries

# Afterword

If the underlying thesis of this book proves to portend the future, density of urban environments will increase, resulting in more juxtapositions of new and old building forms. However, now at the end of the first decade of the third millennium, architecture seems to be at a crossroads where the directions on academic and professional signposts do not align.

In the past few years, the sustainability movement in architectural design has become mainstream. Both students and practitioners are eager to deliver sustainable designs, and more clients are demanding them. Repurposing of old buildings plays an important part in this effort, and historic preservation advocates are well positioned to both foster and hinder the need to save and alter existing structures.

The most widely published designs, however, seem to be on a different trajectory, primarily focused on digitally inspired formal expression. Although many cutting-edge forms that have emerged recently are mathematically inspired, they are not necessarily efficient with respect to energy use.

In practice, the pressure to lower construction costs will continue to affect design quality as the worldwide economic downturn plays out. There may be fewer high-profile commissions as developers and owners consider more modest upgrades to existing facilities in lieu of erecting new buildings from scratch.

These signposts—sustainability/preservation, digitally inspired design, and cost control—seem to be pointing in different directions. The big question for architecture in the next decade will be answered when building designs attempt to resolve all of these forces. Will this be possible?

Existing buildings will endure, especially the good ones, and strong architects will propose exciting new designs that when linked to these structures will enhance our appreciation of both new and old. Controversy will surround many projects as architects grapple with the above issues, but as the case studies chosen for this book illustrate, meaningful architecture can emerge in this genre.

Except for the retention of existing building fabric, the overlay of sustainability has yet to be expressed in work where new and old architecture interact. Soon, perhaps, renewable energy generators like photovoltaics and wind turbines will become a new kind of architectural ornament. This is the next step for both new construction and the hybrid blends of new and old discussed in this book.

Historic preservationists will continue to play an important role saving buildings that deserve to be saved, much like environmentalists working to prevent species extinction. All architects should join this cause and become advocates for saving exemplary buildings of the past.

Although complete alignment of the signposts may not be entirely possible, some very interesting work is certain to appear as future architects attempt to do so. The visual landscape of the future may turn out to be untidy, but it will be diverse to be sure. Overall, this is a good thing—diversity is the planet's very life blood—even in architecture.

# Notes

## Old Buildings

1. Victor Hugo, *The Hunchback of Notre-Dame*, trans. Walter J. Cobb (New York: The New American Library, 1965), 108. Originally published as *Notre-Dame de Paris* in 1831.

2. The Institute of Classical Architecture & Classical America (www.classicist.org) is "dedicated to advancing the practice of the classical tradition in architecture, urbanism, and the allied arts."

3. Adolf Loos, "Ornament and Crime," in *Adolf Loos*, by Panayotis Tournikiotis (New York: Princeton Architectural Press, 1994).

4. Contradiction as a valid architectural tenet was first suggested by Robert Venturi in his famous treatise, *Complexity and Contradiction in Architecture*, published by New York's Museum of Modern Art in 1966.

5. Building codes in the United States define a high-rise building as a structure where the highest floor of occupancy is more than seventy-five feet above the ground. In some jurisdictions, an occupied roof deck can be deemed the highest floor of occupancy. Seventy-five feet is the effective limit of a hook-and-ladder truck parked in front of the building.

6. Responsibility for preservation of historic structures in the United States lies within the National Park Service under the umbrella of the Secretary of the Interior. The "Secretary of the Interior's Standards for Rehabilitation" (www.nps.gov/hps/TPS/tax/rhb/stand.htm) consists of ten broad statements that can be widely interpreted.

Chapter 2
## Sustainable Urban Environments

1. Excerpt from the "Smart Growth Overview" as stated on the Smart Growth Network's website. See www.smartgrowth.org.

2. For a typical example, in San Francisco, a particular six-story building in the downtown core sold at the height of the commercial real estate bubble for over $300 per square foot. In 2009, the building was repossessed; it finally sold in 2010 for under $100 per square foot.

3. The U.S. Bureau of Reclamation was established in 1902 to manage water use as a natural resource. The bureau was responsible for a number of major projects that brought water to land that had suffered from erosion, allowing it to become "reclaimed" for agricultural use.

4. *Tabula rasa* means "blank slate." For artists and architects, the blank slate allows unbiased conceptual design, free of constraints.

5. The mayors of a few major cities in the United States have pledged to support the Kyoto Protocol which in part leads to smart-growth policies, despite the failure of the federal government to do so. The list of these cities includes Seattle, Portland, San Francisco, and Denver.

6. In North America, cities recognized as both dense and livable include New York, San Francisco, Chicago, Boston, and Vancouver.

7. A number of financial incentives, ranging from federal tax credits to local reduction in property taxes, exist to encourage preservation of historic buildings. Most preservation advocacy groups and planning agencies are aware of these incentives and can help owners utilize these benefits.

8. Most major cities have land use policies that allow transfer of development rights (TDRs) from historic properties to other sites. Typically, a historic building may sell the airspace it has not developed within the allowed zoning envelope (e.g., a six-story building in a zone allowing a nine-story building may sell three stories worth of TDRs, usually valued per square foot). The purchaser of the TDRs may apply them to building above the allowed zoning envelope, within special code-prescribed limits.

9. LEED is an acronym for Leadership in Energy and Environmental Design, a point system administered by the nonprofit United States Green Building Council, which certifies sustainable designs. LEED certification has become the benchmark for judging building energy performance.

## Chapter 3
# Design Propositions

1. John Ruskin, *The Seven Lamps of Architecture* (New York: Dover Publications, 1989), 194.

2. Eugène-Emmanuel Viollet-le-Duc, *The Foundations of Architecture*, trans. Kenneth D. Whitehead (New York: George Braziller, 1990), 195.

3. Many historic restorations include modern materials that have replaced archaic materials (e.g., glass fiber reinforced concrete (GFRC) as a substitute for stone). Preservation guidelines, including the "Secretary of Interior's Standards for Rehabilitation," allow this practice, recognizing that exact replication of historic materials, means, and methods would make restoration costs prohibitive.

4. Historic preservationists eschew the duplication of existing work because it leads to a sense of false historicism, blurring the true history of a site. Conservative preservationists look for new work to be in scale and character with the old but of its own time.

5. See pages 11–12.

6. There are almost no major books on the topic of architectural additions, except for one brilliant example, *The Architecture of Additions: Design and Regulation*, by Paul Spencer Byard, FAIA, published by W. W. Norton in 1998. Byard was the director of the Historic Preservation program at Columbia University; his book includes examples of major additions to buildings by well-known architects.

## Chapter 4
# Project Execution

1. Not all owners are bottom-line driven. Institutions and philanthropic entities may have loftier goals that temper the desire to build at the least cost. Almost all owners, however, are concerned about building value, and most often the best architectural solution is not the least expensive one.

2. Value engineering is a process that construction managers advocate so that owners receive the best possible value for a given design intent. In theory, it is an effective process to lower cost without diminishing design integrity by optimizing construction means and methods and finding equivalent but less expensive materials. In practice, value engineering often leads to lower cost by reducing the quality of the design.

3. Phrases that include the word "grandfather" are not usually found in planning or building codes. In general, most codes contain language that will allow existing conditions to persist in buildings that do not conform with current codes if the structures were legally built in conformance with the code of the time. Conditions that were not legal at the time of construction do not become legal over time, a common misperception of the grandfather concept.

4. Shotcrete, or gunite, is pneumatically placed concrete that is sprayed onto an existing surface without the need for formwork. The existing surface can be a soil embankment, a wall, or a single-sided temporary form. The new construction includes steel reinforcing bars and is equivalent to conventional reinforced concrete. Shotcrete is frequently used to strengthen old structures.

5. The Knocktopher Friary and Morgan Library are exemplars of the use of a transition material between new and old—see these two case studies in chapter 5.

6. Assembly of the tower crane for a recently completed project by the author's firm was especially difficult since all of the streets bounding the site contained electrified overhead trolley wires for the city's transit system. The wires needed to be removed temporarily to allow the crane to be assembled on the street and then erected—a two-day process. A major transit artery was shut down over a long weekend, which required careful scheduling and special fees. The process was repeated when the crane was dismantled.

# Chapter 5
# Case Studies

*Dovecote Studio*, Haworth Tompkins
1. Crystal Bennes, "The Dovecote Studio, Snape Maltings, Suffolk by Haworth Tompkins," *Architects' Journal* (January 21, 2010).

*Bar Guru Bar*, KLab Architects
1. Sarah Housely, "Bar Guru Bar by KLab Architecture," Dezeen (September 21, 2009). www.dezeen.com/2009/09/21/bar-guru-bar-by-klab-architecture

*Ozuluama Penthouse*, Architects Collective, at. 103
1. Gabriela Jauregui, "Archi-ology," *Architektur.Aktuell: The Art of Building* (October 2008).
2. Matylda Krzykowski, "Origami by Architects Collective," *Dezeen* (November 16, 2008). www.dezeen.com/2008/11/16/origami-by-architects-collective

*Il Forte di Fortezza*, Markus Scherer and Walter Dietl
1. Catherine Slessor, "Il Forte di Fortezza," *Architectural Review* (April 2010).
2. Ibid.

*Knocktopher Friary*, ODOS Architects
1. Rob Gregory and Catherine Slessor, "Emerging Architecture Awards," *Architectural Review* (December 2009).
2. Jan Vanecek, "Knocktopher Friary," *Archiweb* (May 27, 2009). www.archiweb.cz/buildings.php?type=20&action=show&id=2050

*Walden Studios*, Jensen & Macy Architects
1. At the time of this writing, there were eighteen pieces at Oliver Ranch by the following artists: Terry Allen, Miroslaw Balka, Roger Berry, Ellen Driscoll, Bill Fontana, Andy Goldsworthy, Ann Hamilton, Kristin Jones/Andrew Ginzel, Dennis Leon, Jim Melchert, Bruce Nauman, Martin Puryear, David Rabinowitch/Jim Jennings, Fred Sandback, Richard Serra, Judith Shea, Robert Stackhouse, and Ursula von Rydingsvard.
2. Oliver Ranch, www.oliverranchfoundation.org/about
3. Louise Levathes, "Design for a Floodplain," *Landscape Architecture* (December 2008).
4. Ibid.

*Contemporary Jewish Museum*, Studio Daniel Libeskind
1. Edward Rothstein, "Museum's Vision: West Coast Paradise," *New York Times* (June 9, 2008).

*Morgan Library*, Renzo Piano Building Workshop
1. Victoria Newhouse, "Renzo Piano Alters the Character of New York's Morgan Library and Museum with a New Entrance and Skylit Court," *Architectural Record* (October 2006).

2. Nicolai Ouroussoff, "Renzo Piano's Expansion of the Morgan Library Transforms a World of Robber Barons and Scholars," *New York Times* (April 10, 2006).

*Moderna Museet Malmö*, Tham & Videgård Arkitekter
1. Nico Saieh, "Moderna Museet Malmö/Tham & Videgård Arkitekter," *ArchDaily* (April 2010). www.archdaily.com
2. Catherine Slessor, "Tham & Videgård Arkitekter," *Architectural Review* (March 2010).

*CaixaForum Madrid*, Herzog & de Meuron
1. "Herzog & de Meuron: CaixaForum Madrid," *Arcspace* (March 31, 2008). www.arcspace.com/architects/herzog_meuron/caixa/caixa.html

*1 Kearny Street*, Office of Charles Bloszies
1. John King, "Shared Spirit in 1 Kearny's Styles from 3 Eras," *San Francisco Chronicle* (November 10, 2009).

*Hearst Tower*, Foster + Partners
1. Nicolai Ouroussoff, "Norman Foster's New Hearst Tower Rises from Its 1928 Base," *New York Times* (June 9, 2006).
2. David W. Dunlap, "Landmarks Group Approves Bold Plan for Hearst Tower," *New York Times* (November 28, 2001).
3. Ouroussoff, "Norman Foster's New Hearst Tower Rises from Its 1928 Base."

*Village Street Live-Work*, Santos Prescott & Associates
1. Rachel Strutt, "A Lauded Architect Rescues a Dilapidated Factory from Obscurity," *Boston* magazine, Winter 2009.
2. Bruce Prescott, Santos Prescott & Associates.

*185 Post Street*, Brand + Allen Architects
1. John King, "185 Post Gets the Museum Treatment," *San Francisco Chronicle* (October 28, 2008).

# Acknowledgments

Writing a book is very much like designing a building—it starts with an idea and unfolds in phases with contributions from many people. I am indebted to my entire office, especially Associate Matt Jasmin who urged me to write this book in the first place. Lily Good did a large part of the research, finding some of the best case studies and images, and worked enthusiastically on the myriad details of the manuscript. She and Susan Park reviewed and edited numerous drafts, improving the writing at each step. Katy Hawkins and Katie Crepeau helped develop some of the initial thoughts and were great sounding boards as these concepts developed.

Tim Culvahouse, veteran writer and editor, introduced me to Princeton Architectural Press, and, along with David Meckel, offered advice and encouragement from start to finish. Patrick Bell convinced me that the endeavor could actually be undertaken in parallel with running an architectural practice.

I have drawn from what I learned as a student at the University of Pennsylvania, especially from Professor Holmes Perkins, who instilled in me an appreciation for architectural history, and Professors Robert LeRicolais and Peter McCleary, who inspired my interest in the avant-garde. As a young practitioner, I have Nick Gianopulos to thank for a good professional foundation including a passion for understanding how buildings, both new and old, go together.

Numerous colleagues contributed ideas and examples in response to outreach made for case studies. George Skarmeas recommended that I include Renzo Piano's Morgan Library, which is one of the very best examples of interaction between new and old architecture. Mitchell Schwarzer brought the Duomo at Siracusa to my attention, probably the oldest incidence of architectural styles from different times fusing together.

The architects and photographers whose work appears in this book have provided me with exemplary projects of all types in locations all over the globe. The most enjoyable part of the research was finding little-known work that reinforced the idea of how a union of new and old design can lead to truly good architecture. I am grateful to all of these accomplished practitioners who produced these great designs.

Clare Jacobson was my advocate as she and the editorial staff of Princeton Architectural Press evaluated the initial proposal. Editors Becca Casbon and Megan Carey have guided me through the writing and editing process. Designer Jan Haux has captured the essence of the book's concepts with his precise and well-tailored design. Working with all of them has been terrific. I had no grasp on how difficult it would be to transform an idea into written words.

I now understand why writers always thank their families. My wife, Courtney Broaddus, an accomplished writer and editor herself, helped me move beyond writer's block making a number of key suggestions at critical times. Even the curiosity of our sons, Christopher and Clay, who also saw the Lamborghini in the piazza, helped me to think more clearly about why we react to certain forms the way we do.